POMODORO!

Arts and Traditions of the Table: Perspectives on Culinary History

Arts and Traditions of the Table: Perspectives on Culinary History
Albert Sonnenfeld, Series Editor

A HISTORY OF THE TOMATO IN ITALY

POMODORO!

DAVID GENTILCORE

Columbia University Press ✦ NEW YORK

Columbia University Press

Publishers Since 1893

New York   Chichester, West Sussex

Copyright © 2010 Columbia University Press

All rights reserved

Library of Congress Cataloging-in-Publication Data

Gentilcore, David.

Pomodoro! : a history of the tomato in Italy / David Gentilcore.

p. cm. — (Arts and traditions of the table : perspectives on culinary history)

Includes bibliographical references and index.

ISBN 978-0-231-15206-8 (cloth : alk. paper)

ISBN 978-0-231-52550-3 (e-book)

1. Cookery (Tomatoes). 2. Tomatoes—Italy. 3. Tomatoes—History. I. Title.

TX803.T6G46   2010

641.6'56420945—dc22   2009052786

Columbia University Press books are printed on permanent and durable acid-free paper.

This book is printed on paper with recycled content.

Printed in the United States of America

c 10 9 8 7 6 5 4 3 2 1

For Mum, who still bottles her own

Contents

Preface and Acknowledgments

Italy is Europe's premier tomato nation. Its total production of fresh and processed tomatoes is more than that of all the continent's major tomato-growing countries put together. In Italy today, something like 32,000 acres are dedicated to tomato cultivation, producing around 6.6 million tons of tomatoes for the food industry. The market is worth an astonishing $2.2 billion. Tomatoes are consumed both fresh (raw and cooked) and preserved (in cans, jars, and tubes), and their use in sauces with pasta has become a stereotypical element of Italian cookery, even for Italians. In Italy and beyond, the health benefits of tomatoes are praised as a basic element in the "Mediterranean diet." Lycopene, an antioxidant that tomatoes contain, is known to lower the risk of heart disease, cancer, and premature aging. Never has the tomato enjoyed such favor as it does today.

It took a very long time for this to happen, three centuries in fact. The initial reception of tomatoes was hostile, as sixteenth-century physicians were unanimous in regarding this Mexican native as poisonous, the generator of "melancholic humors." And if the tomato is stereotypical of Italy now, correctly or not, it certainly was not in previous centuries. But there were other stereotypes. In a 1615 letter to the French queen Maria de' Medici from Mantua, the comic actor and playwright Tristano Martinelli wrote that he was as eager to visit her in Paris as "a Florentine is to eat little fish from the Arno, a Venetian to eat oysters, a Neapolitan broccoli, a Sicilian macaroni, a Genoese *gatafura* [a cheese-based pie], a Cremonese beans, a Milanese tripe." Three-quarters of a century later, the same regional stereotypes, plus a few more, were represented in print by Giuseppe Maria Mitelli (figure 1). Not a tomato in sight!

More than three hundred years separate the tomato's first arrival in Italy in the mid-sixteenth century from its consumption and cultivation on a large scale. Even today, there are areas of Italy where relatively few tomatoes are consumed and cultivated.

The combination of pasta with tomato sauce, *pasta al pomodoro*, is so widely consumed that it is hard to imagine that it dates back to only the late nineteenth century, by coincidence around the same time that millions of Italians started crossing the ocean to the New World, where the

Figure 1 Giuseppe Maria Mitelli's print depicts early Italian food stereotypes in the form of a board game, *Gioco della cuccagna che mai si perde e sempre si guadagna* (1691). (© Trustees of the British Museum)

tomato originated. Without this amazing, fortuitous creation, the world would be a much less enjoyable place. Recalling a meal with the dramatist and all-round Fascist celebrity Gabriele D'Annunzio in the 1930s, the poet Umberto Saba recalled the revelation when he was first offered a plate of pasta with tomato sauce. D'Annunzio may have been a disappointment, Saba wrote, but the pasta was not; it was a "crimson marvel" (*purpurea meraviglia*).

How the tomato came to dominate Italian cookery after such inauspicious beginnings is the theme of this book. We shall look at why tomatoes took so long to be adopted and where in Italy they eventually became popular. We shall consider the international nature of the "Italian" tomato, since it traveled across the Atlantic—several times—and beyond. We will look at the presence of the tomato in elite and peasant culture, in family cookbooks and kitchen account books, in travelers' reports, and in Italian art, literature, and film. The story will take us from the tomato as a botanical curiosity (in the sixteenth century) to changing attitudes toward vegetables (in the seventeenth and eighteenth centuries); from the tomato's gradual adoption as a condiment (in the eighteenth century) to its widespread cultivation for canning and concentrate and its happy marriage with factory-produced pasta (both in the late nineteenth century); and from its adoption as a national symbol, both by Italian emigrants abroad and during the Fascist period, to its spread throughout the peninsula (in the twentieth century).

But this is more than just a history of the tomato. From the start, the tomato was closely linked to food ideas and habits as well as with other foodstuffs, with pizza and pasta being only the most obvious. Finally, the tomato's uses were continually subject to change, from production to exchange, distribution, and consumption. For all these reasons, the tomato is an ideal basis for examining the prevailing values, beliefs, conditions, and structures in the society of which it was a part and how they changed over several centuries.

The recipes reproduced in the book are historical in that they originally appeared in printed and manuscript sources throughout the tomato's

Italian history. Although I cannot vouch for them personally, they do not seem dangerous.

This book is part of my ongoing research project on the reception and assimilation of New World plants into Italy, supported by the generous funding of the United Kingdom's Leverhulme Trust. I was able to begin the project in 2005/2006 by taking a year's leave from teaching and administration, which was funded by the Wellcome Trust, as part of a strategic award held jointly by the Universities of Leicester and Warwick, and by a semester of university study leave in 2007. As a visiting professor for six months in 2006, I also was fortunate to benefit from the unparalleled facilities at Villa I Tatti, the Harvard University Center for Italian Renaissance Studies, in Florence. I would like to thank all these institutions for their generous support to make my research and writing possible. I also must acknowledge the interlibrary loan department at the University of Leicester Main Library for patiently and expertly handling a seemingly never-ending series of bizarre tomato-oriented requests, and to the Biblioteca dei Georgofili in Florence, where I spent many happy hours immersed in old farming manuals.

As always, my greatest pleasure has been the interaction with other scholars and interested individuals, especially since I am something of an interloper in the world of food history. I would particularly like to thank Allen Grieco (Villa I Tatti, Florence) for reminding me to pay close attention to details like shape and color when reading historical references; Sheila Barker (Medici Archive Project, Florence) for introducing me to the project's magnificent database; Ken Albala (University of the Pacific) for his infectious enthusiasm for all things related to food and its history; Amy Goldman (New York) for her queries and expertise regarding "heirloom" tomatoes; Alessandra Guigoni (University of Cagliari) for sharing her knowledge and publications concerning Sardinia; and Donna Gabaccia (University of Minnesota) for sharing her work in progress on the Columbian exchange.

I also benefited greatly from the suggestions of my colleague Chris Dyer (University of Leicester), as well as Mauro Ambrosoli (University of Udine), Ottavia Niccoli (University of Trento), Alessandro Pastore (University of Verona), Lucia Tongiorgi Tomasi (University of Pisa), Jules Janick (Perdue University), Nerida Newbigin (University of Sydney), Lia Markey (Florence and Chicago), Joan Thirsk (University of Oxford), and Simone Cinotto (University of Turin).

I am grateful to those people who so generously made their images available for this book: Paolo Cason (www.paolocason.it/libia) for contributing his pictures of colonial Libya, Alfonso Messina (www.solfano.it/sgammeglia) for the archive photos of Canicattì, and Gianfranco Barbera at the Cooperativa Ortofrutticola of Albenga, home of the revived 'Cuor di bue' tomato. Author Pier Luigi Longarini and publisher Maurizio Silva graciously allowed me to use images from Longarini's *Il passato . . . del pomodoro* (Parma: Silva, 1998), and local historian Catherine Tripalin Murray kindly allowed me to reproduce an old photograph from *A Taste of Memories from the Old "Bush": Italian Recipes and Fond Memories from People Who Lived in Madison's Greenbush District, 1900–1960* (Madison, Wis.: Italian-American Women's Mutual Society, 1988).

Finally, I would like to express my thanks to the anonymous readers who read the book on behalf of Columbia University Press, and especially to Jennifer Crewe for her interest in the project and to copy editor Margaret Yamashita for casting an attentive eye over my wordy enthusiasm.

All translations are my own, unless otherwise indicated.

POMODORO!

Modern ITALY
Showing places
mentioned in the text

1 Varese
Sesto S. Giovanni
Milan
Turin
2
Savona
Genoa · Felino
7
Albenga
Imperia
La Spezia
Lucca
Pisa Florence
Livorno
Volterra
Siena
9

Brescia
Varese
Cremona Verona Vicenza Venice Udine Gorizia
Piacenza Mantua Padua Trieste
Parma Colorno
Reggio Emilia Ferrara
Modena Bologna Ravenna
8
Forlì
Castelfiorentino Fano
Arezzo Urbino
Fabriano Ancona

Adriatic Sea

Spoleto Amatrice Pescara
11 10
13
Rome Miglianico
12 S. Marco in Lamis
Fondi 14 Foggia
16 Trani Bari
S.M. Capua Vetere Caserta Melfi
Aversa S. Marzano Monopoli
Naples Avigliano Francavilla F. Brindisi
Torre Annunziata Salerno Potenza Lizzano
Amalfi 17 Gallipoli
S. Cataldo

Tyrrhenian Sea

Corigliano
Cosenza
Taverna Ionian
Nicastro 18 Sea
Polistena Catanzaro
Reggio Calabria
Palermo Messina
Trapani 19
Canicattì
Catania
Siracusa
Lampedusa Vittoria

Sassari
Nuoro
20
Cagliari

The Regions

1 Valle d'Aosta	
2 Piemonte (Piedmont)	
3 Lombardia (Lombardy)	
4 Trentino Alto Adige	
5 Veneto	
6 Friuli Venezia Giulia	
7 Liguria	14 Molise
8 Emilia Romagna	15 Puglia
9 Toscana (Tuscany)	16 Campania
10 Marche	17 Basilicata
11 Umbria	18 Calabria
12 Lazio	19 Sicilia (Sicily)
13 Abruzzo	20 Sardinia

100 kilometers
50 miles

David Gentilcore 2009

1

"Strange and Horrible Things"

The recorded history of the tomato in Italy begins on October 31, 1548, on a day when Cosimo de' Medici, the grand duke of Tuscany, was in Pisa along with his household. His house steward presented a basket to "their excellencies" that had been sent to him. He did not know what was inside, only that it came from the grand duke's Florentine estate at Torre del Gallo. "And the basket was opened and they looked at one another with much thoughtfulness." After the event, the house steward wrote to the Medici private secretary to tell him that the basket of tomatoes—*pomidoro*—had arrived safely.

A curious beginning, perhaps, for the tomato in Italy. But this is the first time the strange new fruit was given a name, anywhere in Europe. While "their excellencies" may have marveled at their first sight of tomatoes, there is no suggestion that they did anything else with them; seeing them for the first time was evidently enough. Four years earlier, a fellow Tuscan, Pietro Andrea Mattioli, had suggested that they might be edible.

Mattioli knew a thing or two about plants. A Sienese, Mattioli was one of the most famous physicians of his day and an avid investigator into the medicinal uses of plants, the area of study that was giving rise to the new field of botany. In *I discorsi . . . della medicina materiale* (1544), his commentary on the most famous pharmacological text of antiquity, by Dioscorides, Mattioli mentions that "another species of eggplant has

Figure 2 The eggplant. (From Pietro Andrea Mattioli, *I discorsi . . . della medicina materiale* [Venice: Felice Valgrisi, 1595])

been brought to Italy in our time" (figure 2). He describes its blood red or golden color when mature and the fact that it can be divided into segments and eaten like eggplant—that is, cooked and seasoned with salt, black pepper, and oil. Ten years later (1554), in a revised edition of his book, Mattioli notes that these curious fruits are called *pomi d'oro*, literally "golden fruits," in vernacular Italian. This is the first time the plant was named in print, and it still is what it is called in Italian.*

We might naturally assume that Italians liked tomatoes right away, but in fact quite the opposite is true. The tomato took more than three hundred years to enter Italian mainstream cultivation and consump-

---

* The Tuscan Italian word *pomodoro* was originally spelled *pomo d'oro*, with a variety of plural forms, such as the now standard *pomodori*, as well as *pomi d'oro*, *pomidoro*, and *pomidori*. Regional dialects usually offer variants of this word, such as the Neapolitan *pummarola*. But there are important exceptions. In Piedmont, it is *tomate* or *tomàtica*; in Lombardy, *tomàtesa*; in Liguria, *tomata*; and in Parma, *tomaca*. Only the Sicilian *pumaramuri* resembles the French *pomme d'amour*, but more on this later.

tion. What took so long? Of all the New World plants in Italy—which include maize, potatoes, tobacco, American beans, and chilies—only the potato took as long as the tomato to catch on. In any case, this is the wrong question. The real mystery is why the tomato caught on at all.

Why were tomatoes regarded with such guarded curiosity? Why were they assumed to be a new variety of eggplant but at the same time called "golden fruits"?

In order to explore early Italian reactions to the tomato, we must start at the beginning. Tomatoes were introduced into Europe following the Spanish conquistador Hernán Cortés's conquest of New Spain (Mexico), which lasted from 1519 to 1521. They seem to have originated in the coastal highlands of western South America, for wild tomatoes still grow in the coastal mountains of Peru, Ecuador, and northern Chile. There is no evidence, however, that these small fruits were ever eaten there. Although we do not know exactly how, they migrated to Central America. There the wild tomatoes were domesticated by the Maya, who developed the larger, furrowed fruits. They in turn were readily adopted by the Aztecs, who cultivated them in the southern regions of Mexico, where Cortés first encountered them.

Returning colonists or missionaries probably brought back these new fruits in the form of seeds, and they then reached Italy from Spain, directly or indirectly, as a botanical specimen. In Italy, the tomato quickly became familiar, or was appropriated, we might say. First of all, as we have seen, it was immediately given an Italian name. Italians preferred to call it *pomodoro* rather than to adopt a variant of the Aztec word for it, *tomatl* (as the Spaniards did, calling it *tomate*). Although Mattioli wrote the commentary while he was a practicing physician in Gorizia, some of the plant names he uses refer to his native Tuscany, and this is may be true of *pomodoro*. Indeed, the uninterrupted use of the term *pomodoro* in Italy is testimony to the plant's continued cultivation since the mid-sixteenth century, an experience rivaling only that of Spain's. Second, the tomato quickly found a place in preexisting plant classifications, being recognized as part of the *Solanum* genus, alongside other familiar plants like the egg-plant and belladonna.

The name Italians gave to the Aztec *tomatl* is an indication of the conflicting notions and half-truths surrounding the wondrous recent arrivals from the New World. The *pomo* of *pomo d'oro* was a generic word for soft tree fruits, similar to the way that "apple" was sometimes used in English at that time. This practice followed from Latin usage (the words *pomum* and *malum*). The *oro* (golden) suggests the importance of color as an indicator when virtually nothing was known about the tomato, never mind that it also came in red. At the time, the use of the label *pomo d'oro* was quite widespread. For example, Gianfranco Angelita's treatise *I pomi d'oro* (1607) was not about tomatoes but about figs and melons. The label *I pomi d'oro* also was widely used to indicate citrus fruits, as Giovanni Battista Ferrari's treatise on their cultivation suggests: *Hesperides, sive de malorum aureororum cultura et usu* (1646).

But the expression *pomo d'oro* was more than simply descriptive. The "Hesperides" in Ferrari's book title suggests a literary allusion to antiquity, to the nymphs of Greek mythology, the Hesperides, who lived in a beautiful garden where a tree producing golden apples grew. Applied to the tomato, it suggests the need to go back to ancient roots in order to find analogies to something new, even if these might be slightly tongue-in-cheek. As the English surgeon-herbalist John Gerard explained in *The Herball or Generall Historie of Plantes* (1597), his discussion of the tomato, or "golden apple," "Howbeit there be other golden apples whereof the poets doe fable, growing in the gardens of the daughters of Hesperus, which a dragon was appointed to keepe, who, as they fable, was killed by Hercules."

This new use of the term to indicate the tomato somehow stuck, although it did not displace previous uses. Confusingly—for me at least, although I suspect also for Italians at the time—*pomo d'oro* continued to denote the mythical fruit as well. So when I came across canto 7 of Lorenzo Lippi's mock-heroic poem *Il malmantile racquistato*, I naturally thought the "golden apples" referred to tomatoes. If so, I decided, it must be an early literary reference, since Lippi began writing the poem in the 1640s. (It was not published until 1676, eleven years after Lippi, also a well-known Florentine painter, died.) In the seventh canto, Brunetto, the estranged and poor father of a girl about to be wed, decides to reconcile with her by providing a crate of *pomi d'oro* as her dowry, overturning it on to the floor in a pile at the couple's feet. If these had been tomatoes, Lippi would have given us the first recipe for making tomato purée. But it was not, as the reference was to antiquity, in Lippi's jocular way.

I experienced the same flutter when I first read about a monumental opera entitled *Il pomo d'oro*. This was no joking matter. Composed by Antonio Cesta and first performed at the court in Vienna in 1668, the opera was more than eight hours in length and had a cast of fifty-one singers, plus choruses, and is often described as the epitome of baroque court operas. Needless to say, it is not about a heroic tomato. Rather, the opera refers again to the ancient myth, the "apple of discord" and the judgment of Paris.

Linking the tomato with the golden apples of the Hesperides is one thing, but mistaking it for an eggplant is quite another. How could this have happened? Why did Mattioli, and many other scholars after him, regard the tomato as a variety of eggplant?

The eggplant itself was a relatively new arrival on the European scene, having come from Persia by way of Andalusia, where it had been introduced by the Arabs. Centuries passed before the eggplant was widely accepted in Italy, and even in the nineteenth century, its consumption varied greatly from region to region. (We shall return to it in chapter 2.) But the tomato and the eggplant do have much in common. Botanically, both are members of the *Solanum* genus. Both also are fruits, from a botanical point of view, as is the cucumber and other cucurbits, since their seeds are enveloped in a juicy pulp. Given their culinary uses today, it is useful to regard them as "vegetable fruits." (Next time someone pedantically reminds you that the tomato is a fruit, not a vegetable, you can reply, even more pedantically, that so are eggplants and cucumbers.)

But this does not answer the question about tomatoes and eggplants. To do so, we must step back a bit and consider the nascent science of botany. Mattioli was part of the reevaluation of ancient sources of knowledge, consistent with Renaissance humanism. In medicine, this reevaluation was marked by a return to Hippocrates and his systematic interpreter, Galen, stripped of medieval accretions. In turn, this return to the ancients meant the daunting task of editing and translating the medical literature of antiquity and then the commentaries based on these texts. Commentaries like Mattioli's, on Dioscorides, sought to reconcile new empirical discoveries in a wide range of fields with old systems of thought. The European discovery of the Americas came just as natural philosophy was rediscovering the knowledge of the ancients, through the first printed commentaries on the likes of Dioscorides, Galen, Pliny, and Theophrastus, all of whom were linked to the Mediterranean basin.

The sixteenth century was a period of unprecedented growth in the study of plants. More than 1,000 Italian species of plants, natural and domesticated, were identified, and 127 new species were brought in from abroad. New World discoveries contributed to the growth of scientific botany in Europe, and the first reaction to these new plants was sheer amazement. Indeed, wonder was a key component of the initial European response to the New World, expressed both emotionally and intellectually, but it was wonder mixed with mistrust. Novelty was suspect; hence the desire to make everything reassuringly familiar. Faced with an abundance still unknown, Europeans sought to reduce the new arrivals to their own norms, to understand and compare them with the familiar plants of Europe, and to look for them in the texts of the ancient authorities.

The conviction still lingered that nothing on earth was unknown to the Bible or the ancients. As a result, investigators like Mattioli saw the new arrivals as variants of those already described by the ancients and thus looked for similarities and analogies. Accordingly, Mattioli was more concerned with "perfecting" the work of Dioscorides, precisely identifying the (mainly Mediterranean) plants it contained, than with exploring the plants of the New World as such. The latter were necessarily relegated to a preexisting unitary vision of nature.

The New World had the potential to lead investigators to call into question the entire cultural system on which their naturalistic knowledge was based. Partly as a result of this, only relatively few of the many new species found their way into printed herbals and botanical works. This absence is curious, since by the end of the eighteenth century most of them (maize, potatoes, manioc, sweet potatoes, various American beans, chilies, peanuts, sunflowers, tomatoes, American squash [*zucca*], cocoa, and vanilla) had found their way into the European diet, with the exception of peanuts and sunflowers, which were not widely consumed until later, and manioc and sweet potatoes, which never have had a significant presence in Europe. Moreover, those species that do appear in the herbals are frequently given an Old World origin, either North Africa or the Middle or Far East. We must remember that the New World continued to be perceived as a part of the Indies and that classical descriptions of plants, like those of Dioscorides, were often so vague that American plants could easily be identified with them. As a result, any "exotic" plant was just as likely to be labeled "Moorish," "Turkish," or "Indian," despite its New World origins (figure 3).

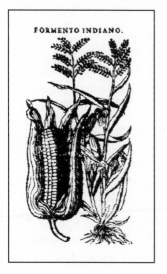

Figure 3 Maize, known in Italian today as *granturco*, or "Turkish wheat." (From Pietro Andrea Mattioli, *I discorsi . . . della medicina materiale* [Venice: Felice Valgrisi, 1595])

Accompanying this mixture of wonder and suspicion was a need to explore, understand, and classify this wealth of new data. Natural philosophers, physicians, botanists, and herbalists eagerly traded seeds and information. This spirit of inquiry and free exchange was not unlike that of today's "seed savers," collectors of so-called heirloom plant varieties shunned by modern intensive agriculture and agribusinesses. To meet their needs, they also required, and produced, a higher standard of botanical description and illustration than that of the medieval herbalists. The study of plants was included in Italian medical faculties. Gardens of "simples," or medicinal plants, ancestors of our botanical gardens, were established, all to facilitate the exchange of information and specimens at the European level. Accordingly, the investigators required a new framework for all this new information. How to classify it? They were limited by the nomenclature then used by scholars throughout Europe, which were merely popular or vernacular names translated into Latin.

Compare Mattioli's response to this situation with that of his contemporary and fellow Tuscan Andrea Cesalpino. Mattioli excelled as an accumulator of botanical information, which he added to in each successive edition of his commentary. But Cesalpino went a step further and sought to systematize this information. As early as the 1560s, the Aretine Cesalpino developed the first binomial nomenclature (genus and species) based on plant characteristics (essentially reproductive characteristics, like fruits and seeds), rather than on properties or uses. There was no love lost between the two Tuscans. Mattioli's commentary went into numerous, lavishly illustrated editions, whereas neither Cesalpino's treatise—which was eventually published as *De plantis libri xvi* in 1583, but without the all-important (but costly) illustrations—nor his system had much success at the time. Not until 150 years later did a Swede named Carl von Linné (Carolus Linnaeus) revive it in his *Classes plantarum* (1738), thereby creating the modern system of plant classification.

The confusion and ambivalence surrounding the tomato can be seen in the flurry of botanical study during the second half of the sixteenth century. Pietro Antonio Michiel, a Venetian patrician and botanist, was a great observer of plants and, for a time in the 1550s, assisted the prefect of the newly founded botanical garden in Padua. Besides cultivating his own garden on the island of San Trovaso in the Venetian lagoon, Michiel was an assiduous trader in botanical news, seeds, and specimens, corresponding with some of the greatest naturalists of the day. Although Michiel accumulated an abundance of classical, Italian, and foreign plant names under the assumption that they would help identify them, his five-book herbal manuscript describing 1,028 plants (of around 730 species), did nothing to resolve any of the confusion.

Like Mattioli, Michiel classified the tomato as a variety of eggplant, providing all the tomato's other names, from classical associations (*Lycopersicon galeni* [Galen's wolf-peach]) and vernacular terms (*pomodoro*) to names used elsewhere in Europe (*poma amoris* and *poma del Perù*). Michiel was content to record unquestioned the names commonly used. If the ancient attribution of "Galen's wolf-peach" and the Peruvian origins of *poma del Perù* struck Michiel as a something of a contradiction, he overlooked it. For instance, he gave the same ancient, indeed Galenic, attribution to the potato. But even some contemporaries realized that they could not have it both ways. As Costanzo Felici expressed it in 1569: "And some authors refer to [the tomato] as Galen's wolf-peach, but if it is a fruit come from Peru, as others suggest, then it couldn't very easily have been known by the ancients."

A further cause of confusion for contemporaries, and more recently for historians, was the identification of the tomatillo (Mexican ground-cherry or husk tomato) alongside the tomato. Today the tomatillo is considered part of the same botanical *family* as tomatoes, the Solanaceae (referring to toxic or narcotic properties), but belongs to another *genus* entirely (*Physalis ixocarpa* [*physalis* = bladder and *ixocarpa* = sticky fruit]). One of the few sixteenth-century authors to distinguish between the two plants was the Florentine aristocrat Giovanvettorio Soderini, who gave them two separate and quite different entries in his agricultural treatise *Della cultura degli orti e giardini . . . ora per la prima volta pubblicato* (1590s). But even Soderini referred to both plants as *pomi d'oro*. Most writers did not even make the distinction.

The monumental work of the Spanish physician Francisco Hernández is a good example. Hernández was Spain's most important doctor, chamber physician to King Philip II and "first physician of all the Indies." At the king's behest, he set sail for the New World in 1571. Hernández eventually compiled a huge description of the flora and fauna there, in 1576 sending back sixteen folio volumes, which were deposited in the Escorial, Philip's beloved palace. These included a detailed description of the tomato's Mexican origins and its different varieties and their uses. But when the fullest surviving edition of Hernández's work was finally published in Italy in 1658, as *Rerum medicarum Novae Hispaniae thesaurus*, the illustration chosen for his discussion of the *tomatl* was of the tomatillo (figure 4).

The confusion had its origins in the Aztecs' use of both plants and, more generally, in the prevailing European attitudes toward Native American culture. The Aztec word *tomatl* indicated simply something round and plump, to which different prefixes were added: *xitomatl* for the tomato and *miltomatl* for the tomatillo. Europeans, however, picked up only the generic term. For the Aztecs, the tomatillo was more common than the tomato. It was grown alongside maize and was used in sauces with chilies, as Hernández himself describes, essentially today's *salsa verde*. The tomatillo's outer husk dries and opens as the fruit matures inside and turns yellow, purple, or pale green. This process seems to have been particularly repellent to European sensibilities: for Hernández, the splitting of the husk to reveal the fruit inside recalled the female genitalia, something he found "horrible and obscene." Accordingly, for Hernández the tomatillo's "venereal and lascivious" appearance was probably the origin of one of its European names, the *poma amoris*. In turn, this became the standard term in French for a century or more, as the *pomme d'amour* and, in English, the "love apple."

It is sometimes assumed that the label "love apple" must come from the tomato's presumed aphrodisiac qualities. But this is a false etymology

 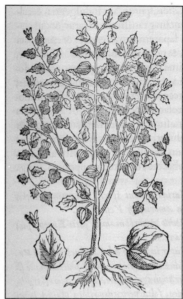

Figure 4 You say tomato; I say tomatillo: frontispiece (*left*) and illustration of the *tomatl* (*right*). (From Francisco Hernández, *Rerum medicarum Novae Hispaniae thesaurus* [Rome: Vitale Mascardi, 1658])

because the tomato had none. For a food to possess aphrodisiac quali-ties in the sixteenth century, it had to be considered "hot" and "moist," according to the Galenic system of food classification. Foods were rated according to their "qualities," which they possessed to greater or lesser "degrees." Hot and moist foods were considered overly nourishing for the average body, leading to "venereal" impulses. Although the tomato was considered "moist," it also was "cold." Far from generating sexual desires, it could be downright harmful. Furthermore, the tomato's asso-ciation with the eggplant and nightshade certainly did it no favors. If eggplants, according to Michiel, "are harmful to the head, generating melancholic humors, cankers, leprosy, oppilations, long-lasting fevers and sickly color," tomatoes "are dangerous and harmful," their odor alone bringing about "eye diseases and headaches." The reason was that they quickly putrefied in the body, causing all sorts of "ill effects." And as if this were not bad enough, the nightshade family included plants like henbane, belladonna, and mandrake, all of which were thought to have potent magical and hallucinatory powers.

Let us return to Hernández. At this time, European scholars were confined to exchanging seeds in order to grow the new American plants and observe them from life, eventually drying and pressing them into large albums. For them, then, plants like the tomato had been uprooted from their original social and cultural contexts, but Hernández, by contrast, was able to observe it in situ. Even so, Hernández's discussion of the tomato tells us as much about prevailing European notions as it does about native Mexican ones. Native American uses had little impact on European perceptions and uses. Because Europeans regarded native societies as inferior, the various roles of their plants were not passed on except in a very superficial way. These feelings of culinary superiority would later come back to haunt Europeans when they came to consume manioc, maize, and potatoes, which at worst can be toxic or at best malnourishing if not prepared in certain ways. In the meantime, their initial encounters with these plants merely confirmed their fear of the new.

Hernández's chapter on the *tomatl* can be found in the section "on sour and acid plants," which already is an indication of how they were perceived and classified by contemporary Europeans. The label *poma amoris* also is European, as we have just seen, as are the disease concepts of the tomato's proposed use as a remedy for eye ulcers, ear tumors, and headaches and stomachaches. A remedy for throat inflammations mixes the juice of the plant with white lead, rose oil, and litharge (lead monoxide), a remedy that has more in common with the chemical medicine of the Swiss physician Paracelsus than it does with the Aztecs.

Europeans considered the New World plants first in medical terms, always hoping to discover the next panacea. This was true whether the writer was a doctor or someone with an eye for trade and profit, as there was money to be made in medicine. The experience with guaiacum, a New World wood used to treat the equally new disease syphilis, was there for all to see. Hernández's suggestions for the tomato, however, never caught on, and its place in the official pharmacopoeias of the seventeenth and eighteenth centuries was never more than a minor one.

Hernández's work nonetheless excited much curiosity among natural philosophers, and some of his original manuscripts were copied for one of Europe's first learned academies, the Rome-based Accademia dei Lincei (Academy of Lynxes). Its founder and sponsor was Federico Cesi, a Roman aristocrat with an interest in botany and natural history. In the first few years of the seventeenth century, Cesi commissioned an illustrated herbal, the *Erbario miniato*, partly based on these manuscripts. The beautiful illustration of the tomato paraphrases the text from Mattioli but adds the word *buoni*, so that, like eggplants, tomatoes are "*good* to eat" (figure 5). No other authority of the time was so favorable.

Figure 5 *Pomi d'oro*. (From Federico Cesi, *Erbario miniato* [early seventeenth century]. Windsor Castle, Royal Library 27713, fol. 23. The Royal Collection © 2009 Her Majesty Queen Elizabeth II)

The tomato depicted in Cesi's herbal is the large, furrowed variety, a cultivated plant not known in the wild state and apparently the variety that the Spanish acquired from the Aztecs. At its best, as in this case, plant illustration became a tool for research, as important as the accompanying text, leading to the symbolic iconography of medieval herbals being replaced by a drive for the most realistic reproduction possible. Only good-quality illustrations could counter the variety of plant names in circulation. By being able to offer the sense of three dimensions, illustrations also had the advantage of being more lifelike than flattened dried specimens. Moreover, artists would depict different varieties of a plant or different stages of growth in a single picture.

Finally in 1628, a summary copy of Hernández's manuscript, the original having been destroyed in a fire, began to see the light of day, in the form of a printed edition. An expanded edition came out in 1658, and only then was it available to a wider educated public. In the meantime, Italian investigators continued with their observation and "experience," based on living specimens, in order to understand their "effects." This was part of a sometimes very slow and painstaking process of identification and verification before the new foodstuffs could enter the food chain. In his influential and successful *Herbario nuovo* (1585), Castore Durante, a physician and lecturer on medicinal plants at Rome's La Sapienza University, described tomatoes as "cold" in terms of Galenic qualities, although not as cold as the fruit of the mandrake. He noted, much as Mattioli had, that tomatoes were eaten like eggplants, with pepper, salt, and oil. But, Durante warned, "they give little and poor nourishment."

From his English vantage point, John Gerard, agreeing that both plant and fruit were cold, went even further, putting the tomato at "perhaps the highest degree of coldnesse." This judgment was based on its moisture and quick growth, features especially pronounced in the English climate, as Gerard found when he tried to grow them from seed he had been sent. For the Paduan physician Giovanni Domenico Sala, writing in 1628, tomatoes and eggplants were among the "strange and horrible things" that "a few unwise people" were willing to eat. The fact that Sala turns to tomatoes and eggplants immediately following a discussion of locust-eating in Ethiopia, spider- and cricket-eating in Padua, and ant- and worm-eating in India highlights the suspicion in which tomatoes continued to be held.

Yet, as both Durante and Sala inadvertently suggest, *someone* in Italy was eating tomatoes, regardless of the dietary advice. Costanzo Felici, a physician near Urbino, tells us who: "gluttons and those eager for new things." We should note that at this time, physicians were increasingly criticizing courtly gluttony and the desire for strange foods. So, as a curious physician interested enough in plants to write at length about them, Felici was just the sort of person we would expect to be "eager for new things." But he is clearly ambivalent about the concept of novelty, and he did not approve of the recipe he had "experienced" either. This involved frying the tomatoes in a pan with the addition of the juice of sour grapes (*uva agresta*), which he described as "to my taste more pretty than good."

From a dietary point of view, as opposed to a strictly botanical one, people like Felici and Durante were asking basic questions of the new plants, according to the historian Madeleine Ferrières: What do you resemble? What do you taste like? What can you replace? The answer to the first question by now should be obvious: the eggplant. As to the second, it is evident that tomatoes, like sour grapes, were appreciated (if that is the right word) for their tartness. Sour grapes were widely used as a condiment, either whole or as juice (called verjuice). It is evident that sweetness had yet to be bred into the tomato; indeed, the tomato that Felici and others were "experiencing" was a long way from most modern varieties. For Felici, the tomato had nothing much to offer other than its bright color. Even the sour grapes were needed to make it tasty. Clearly, the tomato was not ready to replace anything.

In Mattioli's and Durante's simple recipe, the three additional ingredients—black pepper, salt, and olive oil—were meant to correct the coldness of the tomatoes. According to Galenic dietary principles, the qualities of foods had to be balanced, either individually or in combination with other foods. The tomato recipe was European rather than Aztec, and the tomato owed its limited initial use in Italy to the traditional concept of condiment. That is, the condiment counteracted or "corrected" the humoral qualities of the main dish, improving its texture or digestibility. The acidic and cold tomato served to cut through the dryness and heat of chicken and thus remained a "seasoning"—in the literal sense of rendering something else ripe or palatable—for other foods, rather than a food in its own right.

Even when it was eaten, the tomato remained a curiosity, a dietary extravagance, rather than a real foodstuff. Felici's use of the word "experience" is telling in this regard, as it has the sense of testing something unusual or slightly suspect. It is easy for us to laugh at these scholars' quasi-scientific approach to something like the tomato. But in sixteenth-century Europeans' minds, they may have been risking food poisoning. We can only wonder whether the gardeners and assistants and cooks of these elites "eager for new things" were trying them, too.

Just what kinds of tomatoes were the gardeners growing? Often they were the large, furrowed variety just mentioned. But these early Italian botanical and culinary references talk about the tomatillo as much as the tomato. Felici's tomatoes were either yellow and round, and thus perhaps tomatillos, or red and segmented, and thus obviously tomatoes. Soderini describes a variety of *pomodoro* that is definitely a tomatillo, as he says it could be eaten when picked still green, cleaned of its husk (that's the giveaway), sliced, baked in embers or fried in a pan, and salted. By contrast, Soderini notes that the red variety of the plants—our tomatoes—"were to be sought only for their beauty" and were grown only in gardens or flower beds. A decade earlier, Cesalpino had made the same point about tomatoes being grown chiefly as ornamentals.

Tomatoes and tomatillos appear together in a frieze on a large bronze door on the façade of the cathedral of Pisa (figure 6), which

Figure 6 A frieze of tomatoes (and a tomatillo) on a bronze door of the cathedral of Pisa (1600/1601). (Photograph by the author)

was cast in 1600/1601 to replace the doors destroyed in the fire of 1595. The inspiration for the frieze may have come from the borders of Medici tapestries, with their compositions of fruits, flowers, vegetables, and animals entwined around an architectonic backdrop. This also was the time of a renaissance of naturalistic representation in watercolor by artists like Jacopo Lingozzi, based on the direct observation of the natural world and the desire to depict it as realistically as possible.

The Pisan door friezes consist mainly of local flora and fauna, partly hidden in a profusion of foliage, teeming with life. The Christian symbolism of each of the elements depicted on the doors closely echoes various scenes from the Old and New Testaments. Both tomatoes and tomatillos are represented, although they apparently do not have any emblematic significance. Other exotic elements appear as well, all of great naturalistic interest at the time: several sunflowers, a bird of paradise, and a rhinoceros, the last having been an emblem of the Medici family since the time of Alessandro de' Medici. The doors remind us that Pisa's university was well known for its study of the natural world as well as for its botanical garden, founded by Grand Duke Cosimo only a few decades earlier.

In his herbal, Michiel also refers to a variety of yellow "tomato," which may have been a tomatillo that he received from Bologna. He adds, intriguingly, that the yellow variety "hasn't been seen since." Indeed, despite Soderini's preference, the tomatillo inexplicably disappeared from the Italian scene during the seventeenth century, never to reappear.

The tomatillo never became part of the Italian diet, and even now there is no Italian name for the plant. Why this happened is a mystery, for at this time the tomato had no qualities or uses that were evidently superior to those of the tomatillo. As we will see later in our story, the tomato's ability to mutate, so that new and different varieties can be developed and perpetuated with relative ease, contributed to its great success and dissemination. But this feature was not exploited until the nineteenth century, by which time the nonmutating tomatillo had long disappeared from Italy.

Meanwhile, the tomato's curiosity value had extended beyond physicians and other scholars to include wealthy and aristocratic patrons as well as the educated public. These people's gardens mixed botany and pleasure, so just as in the botanical gardens, there was always a place for "exotics." Consequently, this is where the new American plants were first observed and "experienced," and it was in such gardens that tomatoes—like maize, potatoes, tobacco, and American beans—were grown before they were cultivated as crops in open fields. And this point brings us back to Cosimo de' Medici's tomatoes.

Cosimo's Spanish links, through marriage, may have been the source of his tomatoes. His wife, Eleonora di Toledo, owned and managed numerous estates in Tuscany, and she even had produce shipped to Spain. So it is not inconceivable that tomato seeds were an early import. Or Eleonora may have received them from her Spanish father, Pedro de Toledo, who was viceroy of the kingdom of Naples. His own extensive garden and villa were in the Chiaia area of Naples, and Eleonora exchanged gifts of foodstuffs with him. Sharing an interest in New World plants with her husband, in 1545 Eleonora ordered maize (called *grano indiano* [Indian corn]) to be planted at their residence at Villa di Castello. That Eleonora ordered this in November, rather late in the year, suggests that maize was regarded as being like a winter wheat.

Curiosity about American plants was demonstrated in artistic representations as well as in the fields. In 1539, following the arrival of his wife, Eleonora, Cosimo began to turn the Palazzo Vecchio in Florence from an austere medieval residence into a magnificent courtly palace. The large audience chamber was decorated with a series of frescoes by Francesco Salviati, whose battle scenes from the *Triumph of Camillus* (1543–1545) are adorned with garlands of fruits, flowers, and vegetables. Above the sculpted portal depicting Justice is a heavy garland of vegetables, within which we can make out a single ear of maize. And on the next wall to the right, in another garland of vegetables, we can detect two tomatoes, one green and the other greenish red.

The garlands document the range of plants present in Italy at the time, whose functions and uses ranged from culinary and medical to ornamental. The inclusion of plants from the New World in particular served to display the richness and lushness of the natural world, and because they still were rarities, they also showed off the patron's own wealth and power.

The Spanish connection to the New World had to be celebrated. The Medici historian Riguccio Galluzzi reported on Cosimo's attempts to import and acclimatize American plants, and this is supported by Cosimo's private correspondence. In August 1547, Cosimo decided to have a field of maize planted, "of the kind that can be found here in the gardens in and around Florence." Evidently he was not the only one pursuing novelty and did not want to be outdone. He thus ordered his private secretary to procure as much seed as he could and keep it "because he wants to try to sow a field of it and cultivate it to see if it grows." Again, in April 1548, Cosimo ordered a field of maize planted in his garden at Villa di Castello.

Cosimo gave specific instructions that the maize was to be cultivated in the field below the *vivaio*, the villa's specially irrigated nursery and focal point. The entire villa had been renovated in 1537, and Cosimo aimed to turn it into a showpiece, a means of propaganda, and a celebration of grand-ducal power. His intention was to impress visitors to the court, in which the grand duke's botanical collections would have pride of place. The *vivaio*, designed by the Bolognese sculptor Niccolò Tribolo, was a key part of this plan. Tribolo also laid out Pisa's botanical garden, at Cosimo's behest (1545/1546) and the Boboli Gardens. The view was breathtaking. From the *vivaio*, according to Giorgio Vasari, who completed the design, "can be seen in front the palace, the gardens, the fountains, and all the plain below and about them, as far as the ducal villa of Poggio a Caiano, Florence, Prato, Siena, and all that is around for many miles."

Life at the princely courts meant collecting and displaying novelties, curiosities, and wonders from the natural world, of which maize was still one. Maize had been one of the first New World plants that Europeans encountered. It had reached Rome by May 1494 and was being cultivated as a botanical plant in southern Spain by 1535 and in the Veneto region of Italy by 1554, where it was destined to become a staple. Ears of maize were represented artistically as early as 1517, when Giovanni da Udine painted several in a series of festoons for Raphael's frescoes in what is now the Villa Farnesina in Rome. These festoons were the inspiration for the garlands in the Palazzo Vecchio in Florence.

In addition to growing New World plants, Cosimo funded Pisa's botanical garden and lured the famous naturalist Luca Ghini away from Bologna in 1544 to teach there. In 1545, Cosimo rented land in Florence for a "garden of simples," which became permanent five years later. (It is still there.) Cosimo and his successors kept a menagerie as well, containing New World animals like the agouti and a type of weasel. They may have had turkeys, too. The *Primavera* tapestry of 1549, designed by Agnolo Bronzino and now in Florence's Pitti Palace, depicts an American turkey so realistic that it can only have been drawn from life. It is difficult for us to imagine that an animal now so ordinary was at one time a coveted item of courtly exchange. The Medicis' interest in the New World extended to Bernardino de Sahagún's bilingual encyclopedic work on Aztec culture, *Historia general de las cosas de Nueva España* (1576/1577), which was given to Cosimo in 1589, probably by Philip II of Spain. But there is no evidence that its discussion of the Aztecs' culinary use of the tomato directly affected the Medicis' culinary preferences, if they indeed ever read the work. Sahagún's compilation did not find a wider audience until it was finally printed in the nineteenth century.

The passionate curiosity of scholars thus extended to educated amateurs. Sometime before 1547, Lorenzo Cibo Malaspina, duke of Ferentillo, had planted maize at his estate in Agnano (near Pisa), although mice had apparently eaten it. Cibo may have been the person with whom Cosimo was competing. The jurist and aristocrat Francesco Borsati kept his own botanical garden in Mantua, which the Bolognese naturalist Ulisse Aldrovandi hoped to visit it in 1561 on his way to the Council of Trent. When he finally reached Borsati's garden, ten years later in October 1571, Aldrovandi recorded the presence of the tomato—"a yellow, apple-shaped fruit," so maybe it was really a tomatillo—among the more than 350 plants in Borsati's collection. Indeed, Aldrovandi recorded the plant in a separate list of those he desired for his own collection.

The Florentine aristocrat Soderini outshone them all. In the late 1590s, when he was exiled from Florence for his opposition to the Medici, Soderini took up residence near Volterra, where he wrote four manuscript volumes about plants and their cultivation. Soderini's observations are precious because they are based on his own farming experience.

Besides being one of the few to distinguish between tomatillos and tomatoes, Soderini advises how to collect and store the seeds, when to sow
them, and how to treat the tender plants, noting that they require much
manure and water.

Tomato seeds might even be a gift worthy of a king. While traveling through Europe in 1598, the Modenese Giacomo Castelvetro made a
list of seeds to send to Duke Charles in Sweden (soon to be king), with
whom he shared a keen interest in gardening, and on this list was the
tomato. How the seeds fared there is anyone's guess.

The curiosity value of the tomato and other New World plants is
evident in Giuseppe Arcimboldo's imaginative, emblematic, and witty
composite portrait of Emperor Rudolph II, painted around 1590 (figure
7). Arcimboldo depicts the emperor as Vertumnus, the Roman god of
abundance and the changing of the seasons, and represents Rudoph's
accomplishments as ripe fruits, flowers, and vegetables. Two cherry tomatoes form the emperor's lower lip, an ear of maize constitutes his ear, and
two red chilies adorn his cloak.

Rudolph had inherited an extensive garden from his grandfather Ferdinand I. Intended as a "theater of the world," a living encyclopedia of
trees and plants, the garden had been laid out in the Italian style and over-
seen by no less an expert than Mattioli. Sharing his grandfather's fascina-
tion with nature, science, and magic, Rudolph maintained a collection
of "curiosities" from all over the known world, famed for their variety
and value. The Milanese Arcimboldo had been the "court portraitist"
for Rudolph's father and grandfather, although his role at court went far
beyond that. Rudolph entrusted Arcimboldo as his agent, sending him to
Germany in the 1580s to track down works of art and rare objects. Given
all this, it should be no surprise that Arcimboldo would include New
World products in his fanciful portrait. The portrait is similar to the various series of the *Four Seasons* that Arcimboldo had begun to paint almost
three decades earlier, as it brings together all the elements of the *Four Seasons* paintings. By depicting all the fruits, flowers, and vegetables matur-
ing at the same time, Arcimboldo is presenting an allegory of imperial

Figure 7 Lips like cherries: Giuseppe Arcimboldo, *The Emperor Rudolph II as Vertumnus* (1590). (© Skokloster Castle, Sweden; photograph by Samuel Uhrdin)

rule, reminding us of Rudolph's claims of dominion over the world and suggesting the return of a golden age under his rule.

Perhaps, however, the emperor's lower lip is not composed of tomatoes. In a poem accompanying the portrait and describing its contents, Arcimboldo's contemporary Gregorio Comanini does not refer to them. Moreover, the earliest botanic mention that I can find of cherry tomatoes in Italy—indeed, of small tomatoes of any sort—comes fifty years later, even though they most closely resemble tomatoes in their wild form.

Tomatoes, of any shape, hardly ever appear in the Italian art of the time, which is quite remarkable. We would expect to see them in still-life and genre paintings, which overflow with a rich array of fruits, vegetables, and flowers. The tomato's absence from them may be due to its very novelty. At a time when what was depicted on a canvas was conditioned by symbolic or emblematic associations to enrich its meaning, the tomato had none of these.

The tomato appears in only one Italian still life before the mid-eighteenth century, and even that is not certain. This painting, *Flowers, Fruit, Vegetables, and Two Lizards*, in Rome's Galleria Borghese, can be dated no later than 1607, but possibly from the 1580s (figure 8). It is considered one of the first masterpieces of Italian still life, in which a red, ribbed tomato may be lurking between two chilies at the bottom-right-hand corner. Although the art critic Federico Zeri suggested that the painter might have been a young Caravaggio, this identification has been questioned. In any case, the debate need not concern us. What is striking about the painting is the almost "botanical" way in which the fruits and vegetables are arranged and depicted. This almost scientific curiosity, at the expense of a unity of composition, is a common feature of early still lifes. Curiously, some of the objects included—a sprouting onion, a head of celery, scorzonera (a root vegetable), a cardoon, as well as a tomato—immediately disappeared from Italian still lifes, not to return until the eighteenth century.

In the fifty years or so following their appearance in Europe, tomatoes were grown in elite town and country gardens and were occasionally depicted

Figure 8 Master of the Hartford Still-Life, *Flowers, Fruits, Vegetables, and Two Lizards* (before 1607; oil on canvas). (© Galleria Borghese, Rome)

in works of art. But as with Cosimo's tomatoes, we might wonder how often the fruits were eaten. The head gardener at the botanical garden of Aranjuez near Madrid, under the patronage of Philip II, wrote in 1592 that "it is said [tomatoes] are good for sauces." Otherwise, the tomato remained an exotic plant, grown for botanical and ornamental purposes but seldom for culinary uses. The tomato was still more admired than consumed, and although it had been acclimatized in Europe, it had yet to be acculturated.

By the mid-sixteenth century, New World foodstuffs were familiar to learned writers throughout Europe, but it took much longer for them to be used regularly as food. Was this simple dietary conservatism? After all, sixteenth-century Europeans had a built-in reluctance to accept new foodstuffs, based on a mixture of medical notions and feelings of culinary superiority. The rejection was not total, however, and some foods were readily adopted. If the item was analogous to something already eaten, it stood a better chance of being quickly accepted—like maize or beans or chilies, which could easily be classified alongside existing foodstuffs and used in identical or similar ways. The acceptance of the tomato, by contrast, was slow and problematic.

Initial suspicions about the tomato became firmly rooted, limiting its impact. First, in much of Europe, extreme temperatures, insufficient drainage, and inadequate sunlight made tomatoes hard to grow. Second, even in areas where they could grow, like much of Italy, their trailing habit was inauspicious, as proximity to the ground tended to suggest low status. Third, this suggestion was not helped by the presence of both toxic and inedible varieties, which may have taken some time to eliminate. We should remember that even now, all of the tomato plant, with the fortunate exception of the fruit, may be toxic, at least when eaten raw, because of the presence of an alkaloid called tomatine, also found in unripe tomatoes. Fourth, tomatoes had no obvious functions or characteristics to recommend them. Other than color, they had no positive signs for contemporary writers, like flavor or aroma. Even as a simple condiment, tomatoes were regarded as requiring some "help" to make them tasty. Fifth, tomatoes could not satisfy the pangs of hunger; that is, at a time of a largely subsistence diet, they were not as filling as other vegetables already available, so they could not be recommended for the peasant population. Finally, the tomato had no preexisting plant to ease its way into the dietary regimes of the time, and its association with the eggplant, not to mention nightshade, did it more harm than good.

But we must ask whether, in addition to the general problem posed by novelty, there was not something deeper: the widespread European aversion to fruit and vegetables.

2

Death by Vegetables

Can the Italians' aversion to tomatoes be explained by the negative attitude toward fruits and vegetables that prevailed in Renaissance Italy? It was said of Prince Francesco de' Medici, Cosimo's son, that he ate all the wrong foods for a man of his elevated status:

> gross and trivial foods . . . hard to digest, like garlic with black pepper, onions, leeks, shallots, wild garlic, strong raw onions, wild radishes, radishes, horseradish, rampions, artichokes, cardoons, artichoke shoots, celery, arugula, nasturtiums, chestnuts . . . truffles and, in vast quantities, every kind of cheese.

Maybe Francesco's physicians were right. After all, he did die fairly young, not of vegetarianism, though, or even of malaria (as was previously thought), but of arsenic poisoning.

Renaissance physicians lay at the heart of this negative attitude. The dietary advice they gave could not have been clearer. But was anyone paying attention?

Books on health maintenance and hygiene were popular from the sixteenth to the eighteenth century. One of the most successful writers was Tommaso Rangone, a celebrated physician from Ravenna. His manual on "how to live to the age of 120" was first printed in 1550, and to judge by Rangone's medical advice, no one in his right mind would have dared consume vegetables for the various life-threatening infirmities they caused. Foods like vegetables and fruits, perceived as qualitatively watery and viscous, were always vilified, as they got trapped in the body's membranes and putrefied. The moisture had negative effects on the brain, compromising wit and intelligence and, in extreme cases, leading to melancholy. Many vegetables were regarded as suitable only for rustics and laborers, who alone had the bodily heat necessary to counter their cold and moist qualities and the strong stomachs to digest them.

Medical fears of vegetables were occasionally echoed in the culinary literature. The Florentine steward Domenico Romoli ("Il Panonto") concludes his culinary manual, *La singolar dottrina di m. D. Romoli detto il Panonto* (1560), with a "treatise on the maintenance of health." It does not contain a single reference to vegetables. Elsewhere in the book, Romoli advises that "for the maintenance of his health, neither fruits nor plants must a man use as food, because they dampen the humors and do not give nourishment." At best, their cold and humid qualities might be rendered less harmful by being cooked with "hot" ingredients and served on only the hottest of days.

Contemporaries seem to have taken this medical advice with a grain of salt, metaphorically as well as literally. While the success of dietary manuals points to a widespread interest among the elites, satires of the literature suggest the difficulties in following their advice.

Not least of these difficulties were the physicians' contradictory opinions, plus the desire of Renaissance princes and their courtiers to eat what they liked. Because they were rich and leisured, they were the recipients of much of this advice, as it was believed that they alone had the luxury of being able to make dietary choices. But according to the French physician Laurent Joubert, "[Courtiers] never cease interrogating physicians when at the table: Is this good, is this bad or unhealthy? What does this do? Most who ask have no desire to observe what the physician says, but they take pleasure in doing it, for entertainment." Consequently, evidence of princes and courtiers eating all the wrong foods, despite the medical advice, abound.

In fact, edible plants were widely consumed, and not just by Francesco de' Medici. In *Le vinti giornate dell'agricoltura e dei piaceri della villa*, an agricultural treatise printed in 1569, just twenty years after Rangone's manual, Agostino Gallo lists what were then common garden vegetables grown for their usefulness and health-giving properties, which Gallo also details. These were cabbages, leeks, garlic, onions, fennel, carrots, squashes, turnips, radishes, peas, shallots, *erba sana* (allgood, a kind of wild spinach), artichokes, and asparagus. From the point of view of plant husbandry, root vegetables were just as favored as leaf vegetables. Rangone must have been rolling in his grave!

Gallo, a merchant and landowner in Brescia, offers a more accurate snapshot of actual consumption habits than does the physicians' advice. In addition to these common garden vegetables, under "recreational" gardens Gallo lists plants grown for their flavor and for salads (lettuce, radicchio, tarragon, arugula, sorrel, borage, and parsley), soups, and other uses (mint, pennyroyal, Swiss chard, and spinach). Finally, plants were grown in pots to decorate gardens (basil, marjoram, and other "kinds of lovely and sweet-smelling herbs"). None of the plants that Gallo discusses comes from the New World.

Costanzo Felici, a physician and observer of nature with no ax to grind, confirms the impression of a vegetable vogue. Felici notes that the word *insalata* comes from the basic seasoning for edible plants—salt—whether raw or cooked, along with oil and vinegar. The salt serves to dry "the insipid humidity of the herb" and "to give pleasure as well as to prevent it from becoming corrupted and putrefied in the ventricle." He criticizes, however, the way that "salads" are eaten indiscriminately at different times of the day or at different points in the meal, often with no other purpose than to stimulate the appetite so that diners can eat more. Felici is not just referring to elite habits. He also reminds his readers that "toward the end of winter and the beginning of spring, it is said as a proverb among women that any green plant will do for a salad."

The consumption of vegetables was clearly on the rise. The physician Salvatore Massonio reflects this growing passion for edible plants, informing his reader that he has written his erudite *Archidipno* (Beginning of the Meal, 1628) in response to his love of eating salads: "overly pleasing to me and all too frequent." This is a guilty pleasure, as Massonio is bothered that salad, "although so common that it is either eaten or at least known by everyone," is hardly mentioned by ancient writers. Some contemporary physicians continued to have doubts, too. In his advice for magistrates and scholars, Guglielmo Gratarolo commends eating lettuce and endive in salads as "wholesome," but only in hot weather. He admits that colewort and cabbages might be healthy, at least during "cold and moist seasons," although he is worried that the Germans eat them primarily to counteract the effects of too much wine.

In fact, edible plants have a long history as a mainstay of Italian regional diets. In 1596, the English courtier and Italophile Robert Dallington wrote that for poorer Tuscans, "their chiefest food is herbage all the yeare through." "Herbage," he continues,

> is the most generall food of the Tuscan, at whose table a sallet is as ordinary as salt at ours; for being eaten of all sorts of persons, and at all times of the yeare: of the rich because they love to spare; of the poore because they cannot choose; of many Religious because of their vow, of most others because of their want. It remaineth to believe that which themselves confesse; namely, that for every horse-load of flesh eaten, there is ten cart-loades of hearbes and rootes; which also their open markets and private tables doe witnesse.

Dallington's reference to markets is important. The presence of a *piazza delle erbe* in many Italian towns is testimony to the importance of vegetables to both diet and commerce. Savona's "piaza publica de herbe" was where "the most beautiful vegetables" were sold, in the words of a local notary and chronicler, Ottobuono Giordano. In the early sixteenth century, Giordano claimed, with more than a touch of local pride, that the quantity of vegetables sold there was such that "if I went in the morning I would say that they wouldn't finish selling them within a month, but by evening nothing is left. And there are all sorts of vegetables, in both winter and summer." Moreover, this was in Liguria, a part of Italy where cultivating anything meant having to build and maintain terraces in the steep mountainsides, watered by elaborate irrigation systems and carefully fertilized. Nonetheless, the demand for salad was such that Savona's bishop exempted greengrocers and market gardeners from the prohibition against working on Sundays and feast days in order to supply the market with it. Giordano was not exaggerating about the uninterrupted, year-round cultivation of vegetables. At the beginning of the nineteenth century, the same thing impressed the Napoleonic prefect Gilbert Chabrol de Volvic.

Like many Italian towns, Savona had its own trade guild of greengrocers and market gardeners (*ortolani*), and Turin's *ortolani* formed a religious association or confraternity. Some of its members even signed their own names, which was an indication of their rising social status. Gardeners like them would have farmed both inside the town walls and outside in the surrounding countryside. Turin had a network of such gardens, which were sometimes rented out to women. The produce from the market gardens in and around Italian towns formed the bulk of what was sold in their market squares. Thus Goethe noted in May 1787 that "the immediate area around Naples is simply one huge kitchen garden, and it is a delight to see . . . what incredible quantities of vegetables are brought into the city every market day." These gardens were meeting the insatiable demands of what was then one of Europe's largest cities, since Naples was third in size after London and Paris.

By the late sixteenth century, the elites were eating vegetables too, as Dallington suggests. The food of the poor was becoming the food of the rich, even though for the rich, vegetables were only one of a series of dishes served, generally cold. In the early sixteenth century, vegetables were still something of a novelty at court, just beginning to find their way into the myriad dishes, small and large, used to impress visitors. Indeed, the increasing presence of vegetables on elite tables has a whiff of reverse snobbery about it, as Renaissance banquets were scenes of conspicuous consumption and political propaganda as well as occasions for conviviality.

As early as the late fifteenth century, Martino of Como, cook to the Sforza dukes of Milan, included recipes for vegetables in his collection of relatively simple and straightforward dishes. Bartolomeo Sacchi, a cook and humanist par excellence—his status evident in his Latin name, "Platina"—noted the health benefits, or at least how to correct the harmful effects, of a variety of vegetables, such as turnips, spinach, and cabbages.

Perhaps stirred by this advice, in 1519 Isabella d'Este, marquise of Mantua, sent some cabbage seeds to her brother, the duke of Ferrara, "to eat in a salad." These were followed by some actual cabbages "so he can give them a try" (figure 9). Since the duke evidently was unaccustomed to such humble fare, Isabella explained to him that the stems had to be removed first and that the cabbages should be boiled briefly until tender

Figure 9 An oddity fit for a duke: the cabbage. (From Pietro Andrea Mattioli, *I discorsi . . . della medicina materiale* [Venice: Felice Valgrisi, 1595])

and then seasoned with oil and vinegar, "like a salad." "Your Excellency will then see if this oddity is pleasing to him," she concluded.

Edible plants became fashionable at court, especially if they could be presented in elaborate ways (whatever the physicians might say). Giovanni Battista Vigilio's recipe dating from the late sixteenth century, "to make a tasty and lovely salad," consists of fifteen edible plants, seven flowers, nine fruits, and twelve seasonings. Indeed, his idea of a "salad"—as well as the name that he gave to his gossipy chronicle of life at the Gonzaga court—was the "mixing of diverse and various things."

The Italians' consumption of edible plants had regional differences in both caricature and fact. The Lombards were ridiculed as *mangiarape* (turnip eaters), the Cremonese as *mangiafagioli* (bean eaters; but the poor Florentines as *cacafagioli* [bean shitters]), and the Neapolitans as *mangiafoglie* (leaf eaters).

Regarding the "leaf eaters," Emilio Sereni explains that in the Neapolitan dialect of that time, *foglie* referred primarily to broccoli, although not quite the modern variety. The label of "broccoli eaters" was well deserved, as the vegetable was cultivated throughout the year to provide Naples with a steady supply. All parts were eaten—leaves, stems, and flowers—and they were apparently enjoyed by all members of society, from gourmands to paupers, according to a 1646 poem written in praise of broccoli. Not just the Neapolitans loved broccoli. A recipe for "Roman broccoli" (*cavoli alla romanesca*) appears in Martino of Como's recipe collection. This "incomprehensible predilection" for broccoli was shared by the inhabitants of Rome well into the nineteenth century, according to a French physician stationed there. Félix Jacquot referred to broccoli as the "conqueror of pasta" (*le vainqueur du macaroni*) and noted that the broccoli was cooked all day long in vast cauldrons and bought, piping hot, by neighborhood housewives. Both Martino and Jacquot may have been referring, however, to what are today known in Italy as *broccoli romaneschi* (Romanesco broccoli), a light-green cauliflower, combining the form of cauliflower with the sweetness of broccoli.

Neapolitans ate other vegetables too, of course. In 1692/1623, at the Ospedale dei Pellegrini in Naples, a charitable home for pilgrims and the homeless, the basic meal of cabbage, squash, or turnip soup plus salad and fruit was distributed everyday to everyone: staff, *pellegrini*, and convalescents. Large urban religious institutions like the Ospedale dei Pellegrini were great consumers of agricultural produce, as were aristocratic households, and they also were often great producers of it. We know this because they generally kept detailed records, some of which survive for historians to study.

Religious institutions made a virtue of a diet rich in vegetables, so we find nuns eating a large quantity and variety of vegetables in soups, seasoned with *lardo*, which is, strictly speaking, hard bacon fat rather than lard. Olive oil was used more for lighting lamps than for cooking, for which purpose it was reserved for "lean" or meatless days. Later, during the eighteenth century, some vegetables were served raw as a *sopratavola* to nibble on throughout the meal. Fennel, celery, and radishes were eaten in this way. In southern Italy, the Poor Clares of Santa Chiara (Francavilla Fontana, Puglia) made 461 separate vegetable purchases during the year 1748/1749, as did the Capuchin nuns of Santa Maria degli Angeli (Brindisi, Puglia) during 1755/1756. Indeed, in the early 1790s, the nuns of Santa Maria degli Angeli ate a vegetable soup almost daily—made from cauliflower, chicory, cardoons, escarole, or cabbage—as well as vegetables like squash, eggplant, broccoli rabe, artichokes, and *lampaggioli* (a small bitter onion).

The Jesuit priests of Turin were concerned about both the careful administration of their estates and a healthy diet to ensure the accurate record keeping of both production and consumption, and their records detail their increasing cultivation of vegetables and fruit during the eighteenth century. Likewise, aristocratic households combined consumption with production. In areas like Piedmont, Liguria, Lombardy, the Veneto, and Sicily, the second half of the seventeenth and the entire eighteenth century saw the flowering of the aristocratic villa, and the post of head gardener became an important administrative position. Thus the gardener of Vische castle, outside Turin, was reminded that

like a good and diligent father of a family and careful gardener,
[he will] attend to the culture of the gardens . . . that is, for fruit,
flower, citrus, and vegetable [gardens], [he will] tidy and cultivate as
a consequence every fruit tree, whether espalier, bush, or standard,
in the said gardens and nearby vineyards . . . [and] water [them] . . .
to obtain the greatest produce possible from them . . . according to
the best rules of the art.

To northern Europeans, the Italians—like the French—were infamous
for their love of vegetables. Despite an increasing and increasingly varied consumption of vegetables in England, at least by the upper classes,
many English remained wary of "sellets" and other vegetables. This
suspicion could have very real consequences. In 1669, an Englishwoman
took her French husband to court, alleging cruel treatment, which
included his leaving her meatless and very hungry, since he was—to
quote from the wife's deposition—"a Frenchman and useth the diet
of herbs and other slight eating." The salad stereotype persisted for
150 years in the land where eating beef was considered a patriotic duty.
Isaac Cruikshank's print *The Celebrated Mock Italian Song* (1808) ridicules
both the Italian dietary custom and its fashionability in English cof-
feehouses (figure 10):

> With penny-o he will buy any,
> If it have Dandilioni,
> Saladini, beetrootini,
> Endivini, celerini,
> Napkinnini swingidini, . . .

Similarly, Polish satirical poetry of the seventeenth century speaks of
the sons of Polish patricians who go to study in Italy, especially at the university in Padua. But they are soon forced to return, complaining of being
given only salad to eat and never getting a decent portion of meat. Given
this, it is ironic that contact with Italy was crucial to the introduction of

Figure 10 The continuing salad stereotype, in Isaac Cruikshank's *The Celebrated Mock Italian Song* (1808), printed by Laurie & Whittle, London. (© Trustees of the British Museum)

vegetables to the Polish diet. Indeed, the Polish language abounds with Italian terms for vegetables. The words used in Poland for tomato, cauliflower, onion, asparagus, zucchini, and chicory are Italian, as is the general term for greens and soup vegetables, *włoszczyzna* (literally, "Italian things").

According to Polish tradition, vegetables were introduced by Princess Bona Sforza d'Aragona and her retinue when she came from Bari to marry King Sigismund I in 1517. During her four decades in Poland, her cooks supposedly prepared the Italian specialties of the time. But this may be something of a myth, similar to crediting Catherine de' Medici with single-handedly creating modern French cuisine through her Italian influence, following her marriage to Henri II in 1533. In fact, however, Polish commercial and culinary contacts with Italy date back to the fourteenth century.

In any case, national variations in diet were believed to exist for a reason; that is, they were the result of differing climates and constitutions, which one disregarded at one's peril. The "heat" of digestion of northern and southern Europeans differed because their climates differed, and the inhabitants were expected to eat accordingly. Novelty in diet posed hidden dangers. William Harrison, writing in the late sixteenth century, deplored the recent fashion among English merchants and nobles of eating dangerous foods like "verangenes" (eggplants), "as if nature had ordained all for the belly." Such action went against the divine order of things. Only when new plants had become fully acclimatized would they become transmuted into native—and therefore safe—commodities. John Gerard said much the same in regard to eggplants, which never ripened in the English climate: "I rather wish English men to content themselves with the meat and sauce of our owne country, than with fruit and sauce eaten with such peril; for doubtlesse these [eggplants] have a mischievous qualitie, the use whereof is utterly to bee foresaken." Gerard concludes, "It is therefore better to esteem this plant and have it in the garden for your pleasure and the rarenesse thereof, than for any virtue or good qualities yet knowne."

An Italian living in England, Giacomo Castelvetro, had tried to buck this trend. In 1614, he found himself in England on the run from the Inquisition in Italy for his Protestant beliefs. A keen promoter of Italian culture, Castelvetro dedicated his "Brief account of all the roots, plants and fruits that raw or cooked are eaten in Italy" (1614) to the Italophile Lucy Russell, countess of Bedford. Castelvetro's quixotic mission—aside from a search for patronage—was to encourage the English to eat more vegetables and fruit and less meat and fewer overly rich dishes and sweet things. He was not very successful. Although several manuscript copies of the short work survive, it was not translated and published until our own time. Castelvetro did not obtain any patronage, either.

Nevertheless, his essay is a breath of fresh air when compared with contemporary medical-dietary treatises, conservative in their Galenical ideas and heavy on the theorizing. Castelvetro wears his knowledge lightly, but more than that, he is not really concerned with the notion of a close fit between national constitutions and national diets. Rather, what is good for the Italians and the French must also be good for the English. The benefits of vegetables—"they are refreshing, they do not thicken the blood, and above all, they revive the flagging appetite"—applied to all people.

Castelvetro was familiar with the tomato, as we saw in chapter 1, but his interest appears to have been limited to exchanging seeds. At this stage, the tomato was still too new to be included in a treatise on culinary practices, but the eggplant, botanically associated with the tomato, merits an enthusiastic entry in Castelvetro's essay. Evidence of eggplants in Italy is contradictory. Many people were still suspicious of them. Although prevailing medical notions were partly responsible, their association with Italy's Jews was no doubt a factor, too. When they were expelled from Spain, Jews and Jewish converts from Andalusia may have brought their taste for eggplants with them. In addition, many Italian Jews had contacts with the Middle East, where they sometimes sought refuge and where the eggplant originated. Yet it is strange that until the nineteenth century, eggplants are virtually absent from contemporary descriptions, in both Hebrew and Italian, of Italian Jewish dietary habits. Their consumption appears to have been limited to poor Jews. In any case, though, in Florence as late as the 1850s, eggplants were apparently still difficult to find and were disdained by many people as "food of the Jews." By contrast, in southern Italy from at least the end of the eighteenth century, they were being widely consumed, as we shall see later.

In the seventeenth century, the consumption of vegetables rose throughout Europe, and this continued into the following century. New varieties of plants kept up the public's interest. John Evelyn, an English virtuoso and fellow of the Royal Society, wrote in praise of the "wholesomeness of the herby-diet," and Louis Lémery, a French chemical physician, extolled the value of the plant-based diet of primitive society, "when men lived longer and were subject to fewer diseases than we."

Neither writer appears to have been a vegetarian, and, indeed, the word itself was not coined until the mid-nineteenth century. Few writers were willing to recommend a vegetables-only diet, except for certain categories of sick people. Nonetheless, when he was the grand duke of Tuscany, Cosimo III apparently became a complete vegetarian under the advice of his physician, Francesco Redi. Otherwise, physicians tended to scorn the idea that "one can substitute without risk a meager (vegetable) diet for one of flesh," to quote the Frenchman Nicolas Landry.

The Florentine physician Antonio Cocchi disagreed (figure 11). Cocchi's *Del vitto pitagorico* was published in 1743 and, two years later, translated into English as *The Pythagorean Diet, of Vegetables Only, Conducive to the Preservation of Health, and the Cure of Disease*. Cocchi claimed that

Antonio Cocchi

Figure 11 The physician and vegetarian advocate Antonio Cocchi. (From Cocchi, *Discorsi e lettere*, vol. 1 [Milan, 1824])

Tuscans were among the healthiest people in the world. The reason was their poverty, which forced them to rely on vegetables and fruit and eat very little meat (reiterating Dallington's comments of 150 years earlier). Cocchi put his own gloss on the latest physiological understandings of the digestive process to explain how a vegetarian diet would benefit people. Optimal nutrition depended on what Cocchi called "subtlety": the lightness, clarity, and mobility of the body's fluids. A meat-based diet was too dry and difficult to digest, clogging the body's passages; by contrast, fruits and plants provided a more readily abundant and usable form of fluid.

Cocchi turned to the ancient ascetic Pythagoras for his inspiration, and not the "barbaric school" of ancient physicians, who had regarded fruits and vegetables as too watery and phlegmatic (their ideas had been revived in the Renaissance). Cocchi's advice is to avoid all "invigorating and pungent" vegetables, like onions, garlic, and bulbous roots, for the overly "solid nourishment" they provide. Here we find echoes of the ancient distrust of root vegetables. We also must avoid all dry fruits, nuts, and the hardest seeds. That leaves around forty vegetables, which Cocchi does not identify but which are "normally cultivated amongst us in our fields and gardens; and those which are the most common, are also the most wholesome."

Few physicians were willing to go as far as Cocchi. The problem with vegetables, argued Cocchi's contemporary Giuseppe Antonio Pujati, was that they were *too* easy to digest. That is, they were expelled from the body so easily that they provided little nutrition. An exclusively vegetable diet, such as that of the poor forced to subsist on wild plants, resulted not in better health but in intestinal problems, vomiting, and diarrhea. Consequently, Pujati concluded that vegetables should be considered as "correctives" to meat and other dishes and not nourishing foodstuffs in their own right.

The problem was that "a vegetable diet renders man gentle and pleasant in nature," which we might think was good. But someone had to grow these vegetables, and paradoxically, the enforced "Pythagorean regimen" of Italy's peasants made them "weak and incapable of great labors." It was one link in a chain that had resulted in the "decline of agriculture."

Historians always have difficulty finding information relating to the diet of the poor. Luckily, when the kingdom of Naples (southern Italy) came under the rule of Napoleon, the French investigated local economic and social conditions. Their resulting report, the *Statistica murattiana*, covers the years 1807 to 1811. (If it means jumping ahead a few decades in our account, it is certainly worth it, given the source's wealth of information. In any case, most of the investigation's findings can be projected back a generation to the late eighteenth century.) The *Statistica* was based on similar projects elsewhere, following the installation of Napoleonic governments in much of Italy, as well as on inquiries carried out in France. Each of the Neapolitan provinces had a local editor, who usually was a reform-minded physician, lawyer, or agronomist responsible for circulating the questionnaires, assembling the findings, and returning them to Naples. Answers to the questions relating to the local diet and the rural economy provide some fascinating insights into the condition of the local populations throughout southern Italy. Here much of the population not only ate vegetables but even subsisted on them, and here the tomato first became an important part of the local diet.

The editor for the province of Abruzzo Ulteriore remarked that "the tendency toward a vegetable diet depends very often on habit and poverty, which obliges people, reluctantly, to turn to whatever food they can most easily procure." Wild plants were eaten everywhere. In the words of the compiler for the province of Molise,

> Generally, green soup [*la minestra verde*] is the ordinary food of a peasant family. It is often composed of wild edible plants. . . . They are boiled in water, seasoned with oil or pork fat, salt, pepper, or ground chilies. This food is good for health and costs nothing; one girl can gather in three hours in the countryside enough for a family of five.

The editor for Abruzzo Citeriore took a different view, bemoaning the overreliance on "field vegetables," which were not always properly seasoned and contributed to "weakness, cachexia, dropsy, putrid fevers and other epidemic diseases, as was sadly experienced in the year 1803." Most compilers concurred.

In Capitanata, the peasants ate something they "call *acqua e sale* and bread cooked in oil, and very often they mix wild plants in with it." *Acqua-sale* was the name given to a dish eaten in Puglia, Basilicata, and Calabria, in which stale bread was moistened with water and oil, and then mixed with some salt and vegetables like garlic, onion, celery, sweet peppers, and, as they became more widespread, tomatoes.

The *Statistica* gives the impression of a southern Italian peasant diet that is full of variety. But this variety had its limitations. First, it was a diet largely limited to vegetables, accompanied by bread made largely from inferior cereals. Second, as a result, it was also highly seasonal, which explains the importance to the domestic economy of preserving foods for use throughout the year (to which we shall return in chapter 4). Third, given the nature of the source, the *Statistica* tends to record peasants' production rather than consumption. Much of this "variety," therefore, was actually cultivated to be taken to market and sold, not eaten. Finally, methods of preparation remained basic because the kitchen utensils remained basic. The peasant kitchen thus was simple, with only a few clay or wooden implements. In contrast, the better-off kitchen was increasingly stocked with a wide range of utensils for the preparation and preservation of foods.

On a more positive note, the *Statistica* shows traces of an appreciation for vegetables shared throughout society. Francesco Perrini, a cathedral canon and the compiler for the rich agricultural province of Terra di Lavoro, was a real vegetable enthusiast. Perrini goes out of his way to praise the two different varieties of fennel from the towns of Sora and Nola, the onions of Mondragone, the cardoons of Cerreto, the cauliflowers of Matese, the white radishes of Aversa, the ball lettuces of Sant'Angelo, the head cabbages of Santa Maria, and the chilies of Maddaloni. This was the hinterland of Naples, so supplying the city with these products made a significant contribution to the local economy.

Perrini even extensively details how different vegetables were consumed. His list includes important New World products. One of them has been mentioned, the chilies of Maddaloni. The variety grown, *baccato* (*Capsicum baccatum*), produced a "small, red fruit, but of an insufferable pungency," Perrini noted. The chilies were used to season eggplants: "Ordinarily [eggplants] are served boiled, sliced and then cured with the strongest vinegar possible, to which is added a dose of *baccato* chilies, oregano, and slivers of bruised garlic, called by the local name of *impepata* or *molignanelle*." Eggplants were also eaten fried, although the well-off preferred them stuffed. Some varieties of peppers—the long "common" pepper and the yellow bell pepper—also were eaten as vegetables. Perrini describes that they were "cooked in the ashes, raw and fried, or in salads or cured in vinegar so they can be kept for the whole year."

Even the potato had made limited inroads, Perrini was pleased to report, among the inhabitants of the Apennines. These mountain dwellers ate them seasoned "with oil, pork fat, cooked grape must, or salt and prepared different dishes with them." This comment should be seen in the context of a long campaign by reformers throughout Europe, begun in the eighteenth century, to have the potato adopted as a staple crop as a means of combating hunger and famine.

But among all these favored foodstuffs was one plant that, despite having no government backing, was destined to have the greatest impact on the future of Italian cookery: the tomato.

3

"They Are to Be Enjoyed"

A survey on the condition of farming in Tuscany, conducted in 1759 by the physican-botanist Giovanni Targioni Tozzetti, included tomatoes among the "fruits prized by men as foodstuffs or as condiments for them." By this time, the tomato was not just an agricultural and culinary presence but also an artistic presence. Carlo Magini was the first Italian painter since the very early seventeenth century to put tomatoes in a still-life painting, and he did it on numerous occasions. A native of Fano in the Marche, Magini was active in the second half of the eighteenth century, making a modest living by painting portraits, devotional works, and still lifes. His still lifes are simple kitchen scenes, painted in the neoclassical realism of the time and offering the viewer an intimate contemplation of everyday objects that may have reflected the painter's own social status, which was far from well off. Magini represented objects like the clay pots typical of the region, a brass lantern, a bottle of local wine, a loaf of bread, eggs in a pan, cheese on a plate, cured meats, vegetables, and fruit, all usually arranged on a small table or buffet. Each painting was a meal or a dish waiting to be cooked and consumed. Furthermore, Magini's depictions closely fit actual dietary practices. Thus a real-life lunch eaten at the Jesuit teaching college in nearby Loreto on a Friday in January 1720 consisted of "soup of black-eyed beans, portion of eel, pan-fried egg, apple, and cheese."

Figure I 2 Carlo Magini, *Still Life with Mullet, Grill, Wine Flask, and Glass* (second half of eighteenth century; oil on canvas). This is one of the few still-life paintings in which Magini depicted fish, which is strange given that he spent almost all his life in the coastal town of Fano. (Courtesy of the Pinacoteca Comunale di Faenza)

Of the no fewer than ninety-nine still lifes attributed to Magini, six contain tomatoes. They are furrowed and misshapen, in groups of two or three, red and green, and usually in one of the bottom corners. Whether tomatoes were eaten green or were included for chromatic contrast is hard to say. Four of Magini's still-life paintings do, however, suggest how the tomato might have been prepared and eaten: combined with eggplants and squash, to be served with cheese; or as a condiment with eggs and cardoons (*gobbo*), chicken (with celery and onion), or grilled mullet (figure 12). An intriguing possibility is that tomatoes were being eaten in salads, since Magini once places one alongside lettuce and cucumbers. Besides Magini, a contemporary of his, Nicola Levoli, who was active just slightly to the north in Rimini, had a similar background and painted similar scenes. Levoli's *Still Life with Fish on the Grill* (1770–1800) contains three tomatoes, two red and one green.

What had happened since the early seventeenth century to alter attitudes toward the tomato? This chapter looks at how the acclimatization of the tomato slowly gave way to its acculturation during the eighteenth century and how the tomato's "botanical" phase, examined in chapter 1, paved the way for its more widespread cultivation and consumption.

The changing attitude toward fruits and vegetables that characterized the late Renaissance period was partly responsible, as were the changing medical and scientific notions about how the process of digestion worked.

By the mid-seventeenth century, medical ideas had shifted away from the dominant Galenic ideology of previous centuries. Consequently, the system of the "humors" began to lose authority among physicians as well as dietary writers. Moreover, the new medical ideas meant that the tomato, as long as it was cooked, was no longer perceived as a health risk. This shift was characterized by two new medical theories, sometimes competing, sometimes complementary. The first of these was "chemical" medicine based on the ideas of Paracelsus and Jan Baptist van Helmont, which saw the body's functions as chemical processes. The second was "mechanical" medicine, which regarded the body as a machine. According to this new mechanical model of digestion, the body's input and output could be measured and quantified mathematically.

Previously, according to Galenic humoral notions, the tomato was thought to hinder digestion, owing to its acidity and its cold and damp nature, which, Galenic physicians thought, remained in the body and corrupted the humors. Now, however, the tomato's characteristics were believed to help the process by breaking down foods. The mixture of salty and acidic foods seemed to aid in the "fermentation" of food in the stomach. Thus although the tomato itself had not changed, the attitudes toward it had.

Another, broader, point affected ideas about food and diet as well. Partly as a result of these new medical systems, seventeenth- and eighteenth-century physicians were less interested in discussing the relative merits and threats of different foodstuffs. Instead, medicine was increasingly directed to the definition and treatment of diseases, rather than the care of healthy bodies. Dietetics was accordingly separated from medicine and connected with other arts, like cooking. Eventually, by the early nineteenth century, a new field had taken the place left by dietetics: gastronomy, the study of "eating well."

The tomato's fortunes began to shift during the mid-seventeenth century. In 1640, the physician and botanist Pietro Castelli published *Hortus messanensis*, a description of the botanical garden in Messina, Sicily, in which the tomato appears under the heading "Medicinal Plants." Castelli did not think of the tomato as a "garden plant," which is how he classified the eggplant, nor did he list it with the nonmedicinal plants and flowers. But even though he regarded the tomato as a medicinal plant, Castelli must not have thought very much of it, as he did not even mention it in his detailed commentary on the medicinal ingredients found in Rome's official pharmacopoeia.

From other sources, we know that the tomato was being eaten and not just grown. By the 1660s, the tomato appears in a list of the "many fruits [the Italians] eat, which we either have not, or eat not in England." The list was compiled by the theologian, naturalist, and gardener John Ray during a three-year visit to the Continent from 1663 to 1666. In another, later work, Ray notes that in Italy tomatoes are cooked with marrows, pepper, salt, and oil. Although he had doubts about how healthy tomatoes were, Ray did believe that if they were cooked in oil, they might be good for scabies (then a widespread skin complaint). The tomatoes were not to be eaten, though, but rubbed on to the skin to take advantage of their perceived cooling and moistening effects. The remedy was not popular.

In Ray's England, the tomato was a rarity. Still referred to as the "love apple," it was obscure enough to be included in a "dictionary of difficult terms" published in 1677, and even then the dictionary's author managed to get the definition wrong. He referred to it as "a Spanish root of a color near violet," confusing it with both the eggplant and a root vegetable. Then, fifteen years later, the author of *The Ladies Dictionary, Being a General Entertainment of the Fair-Sex* (1694), repeated the errors, word for word.

Ray's new Italian recipe for the tomato may have been inspired by Francisco Hernández's *Rerum medicarum Novae Hispaniae thesaurus*, a description of the plants of New Spain, which finally was published in 1628 and again in 1658, as we saw in chapter 1. Hernández's discussion of the native Mexican use of the tomato, with its combination of ingredients, was new to Italy. "One prepares a delicious dip sauce [*intinctus*]," Hernández writes, "from chopped tomatoes, mixed with chilies, which complements the flavor of almost all dishes and foods and wakens a dull appetite."

The tomato owed its simpler, initial uses in Italy, as described in chapter 1, to the traditional concept of condiment. Condiments and seasonings served to balance a dish's humoral "qualities." By the seventeenth century, these combinations—think of serving prosciutto (salty, dry) with melon (watery, sweet), still a popular combination—were becoming routine. They were popular more for their pleasing gustatory contrasts than as humorally balanced dishes, however. Italian food writers were paying less and less attention to corrective rules in their combinations of foods and more attention to flavor. Personal experience and preferences, local practices and habits, were beginning to take precedence over ancient authority and received wisdom.

This newer association of tomatoes and squash, and sometimes with eggplants and chilies, entered Italy by way of Spain. This must have been a recognizable combination there by the time Bartolomé Murillo painted *The Angels' Kitchen* (1646) for the Franciscan monastery of Seville. The right-hand side of the painting shows angels preparing a meal. In the corner, presented together as ingredients, are a red furrowed tomato, two eggplants, and a squash. Unfortunately, Spanish printed sources—plant books and cookbooks—do little to verify or contextualize this use. But we do know that the Sevillian Hospital de la Sangre made two purchases of tomatoes (and cucumbers) in July and August 1608.

The meal being prepared in *The Angels' Kitchen* was essentially the recipe that the Italians, especially the Neapolitans, then ruled from Madrid, acquired from the Spanish. At least, this is how Antonio Latini acquired it. According to his autobiography, *La vita di uno scalco*, after an unfortunate start as a poor orphan, first in his native Fabriano (Marche) and later on the streets of Rome, in 1658 Latini was taken into the kitchen service of Cardinal Antonio Barberini, one of Pope Urban VIII's numerous nephews. There, Latini quickly rose through the ranks to become kitchen steward (*scalco*). Over the next few years, Latini served as a steward in other noble and ecclesiastical households in Macerata, Bologna, and again in Rome, before being offered the post of steward to Esteban Carrillo y Salcedo, a grandee of Spain and regent to the Spanish viceroy of Naples. This was powerful, wealthy Spain, with an empire bridging the Old and New Worlds, so it was a great success for the forty-year-old Latini. Upon his arrival, Latini was given forty gold *scudi* to defray his travel costs and his new clothes "in the Spanish style" (figure 13). He was now in charge of cooking in Carillo's villa on the slopes of Vesuvius, overlooking the Bay of Naples, where Carillo "often banqueted with the most noble personages in royal splendor and magnificence." Here Latini was rewarded with the titles of knight of the golden spur and count palatine, dictated his autobiography in 1690, and compiled his masterpiece, *Lo scalco alla moderna*, published in two volumes a few years before his death in 1696.

Latini's cooking is at once refined and eclectic, borrowing from his own broad range of experiences and contacts. On the one hand, it is the culmination of Italian court cooking before the triumph of French cuisine in the eighteenth century. On the other hand, Latini is not afraid to use popular food traditions, from vegetable soups to tripe and other offal; to develop a "new way of cooking without spices," using herbs rather than strong flavorings; and to experiment with newer ingredients, like turkey, chocolate, chilies, maize, and, of course, the tomato. All the

Figure 13 Antonio Latini posing in all his Spanish finery. (From Latini, *Lo scalco alla moderna* [Naples: Domenico Antonio Parrino e Michele Mutii, 1692, 1694])

dishes in which the tomato appears are indicated as "in the Spanish style" (*alla spagnuola*). The first is a fiery tomato condiment to accompany boiled foods. The second brings together eggplants, squash, tomatoes, and onions, a combination that became a Mediterranean standby, apparent in the Catalonian *samfaína* and the Provençal ratatouille. The third tomato recipe takes the form of a hearty mixed meat stew, named after the pot, *cassuola*, in which it was cooked. Latini named this dish *cassuola alla spagnola* when he served it as one of fourteen hot first-course dishes at a wedding banquet in April. How he got tomatoes in April, Latini does not say. (This was almost three centuries before the Dutch mastered year-round hot-house production.)

Latini's three recipes are the first time tomatoes were used in European culinary literature. They met the increasing demand for condiments and dishes that were flavorful but not based on spices. Cooks now were

Antonio Latini's Tomato Recipes

Salsa di pomodoro alla spagnola (tomato sauce, Spanish style). Take half a dozen ripe tomatoes and roast them in embers, and when they are charred, carefully remove the skin, and mince them finely with a knife. Add as many onions, finely minced, as desired; chilies [*peparolo*, in Neapolitan dialect], also finely minced; and a small amount of thyme. After mixing everything together, add a little salt, oil, and vinegar as needed. It is a very tasty sauce, for boiled dishes or anything else.

•••••

Minestra alla molignane (eggplant dish). Cut [the eggplants] into small pieces; add minced onions and squash [*cocuzze*], likewise cut small; and diced tomatoes. Lightly sauté everything together with aromatic herbs, with sour grapes if they are in season, and with the usual spices. You will produce a very good dish, Spanish style.

•••••

Cassuola di pomadoro (tomato casserole). Fill the pot [*cassuola*] with pieces of pigeon, veal breast, and stuffed chicken necks. Stew well in some good broth, with suitable aromatic herbs and spices, together with cockscombs and testicles. When the stew is cooked through, roast some tomatoes in embers, peel them, cut them into four pieces, and add them to the soup along with the rest of the ingredients, making sure not to overcook them, as they require little cooking. Then add some fresh eggs and a little lemon juice, and allow the mixture to thicken, covering it with a lid and applying heat both above and below.

Antonio Latini, *Lo scalco alla moderna* (Naples: Domenico Antonio Parrino e Michele Mutii, 1692, 1694), 1:444, 2:55, 1:390.

trying to stimulate the appetite with delicate and pleasing foods. If they became more digestible in the process, so much the better; but it was not their main aim any more.

A recipe collection dating from 1705 suggests how the tomato's use was spreading (figure 14). The manuscript's author, Francesco Gaudentio, was born in Florence in 1648 and became a lay assistant with the Society of

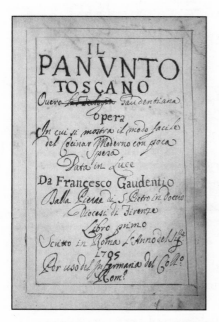

Figure 14 The title page of Francesco Gaudentio's manuscript recipe collection "Il Panunto toscano, overo la teologia gaudentiana ..." (1705). Gaudentio was evidently inspired by fellow Florentine Domenico Romoli, nicknamed "Il Panonto" (after a simple specialty of seasoned, toasted bread), the author of a culinary manual first published in 1560. (Courtesy of the Biblioteca Città di Arezzo)

Jesus (Jesuits). He worked first at their colleges in Spoleto and Arezzo before joining the Collegio Romano, the Jesuits' flagship educational institution in Rome, where he was a porter and cook at the college's infirmary until his death in 1733.

Because the Collegio Romano was residential, food was important. Moreover, as the Catholic world's foremost missionary and teaching order, the Jesuits took food and diet seriously. Each of their institutions kept extensive records of what food was served each day. These records were known as a *Levitico*, from the Levitical laws and rituals, because they detailed the different foods consumed on fast and feast days, according to the Catholic Church's calendar. The *Levitico* for the Roman Province of the Society of Jesus suggests a balanced regime of vegetables, legumes, and fruit; meat and fish; and pasta, bread, and rice, accompanied by cheese and eggs. Supplementary foods were offered to those taking medicines and to visiting guests and dignitaries. This regime is filling and

Francisco Gaudentio's Tomato Recipe

How to cook tomatoes. These fruits are something like apples, as they are cultivated in gardens and are cooked in the following way: Take the tomatoes, cut them into pieces, and place them in a pan with oil, pepper, salt, minced garlic, and mint. Sauté the tomatoes, turning them frequently, and if you wish, add some tender eggplants or squash and they will improve it.

Francesco Gaudentio, "Il Panunto toscano, overo la teologia gaudentiana . . ." (1705), MS 450, fols. 366–67, Biblioteca Città di Arezzo.

nutritious but not extravagant, although, of course, most Italians then could only dream of such a diet. Indeed, roughly one-third of the Collegio Romano's expenses in the late seventeenth century were for food.

Gaudentio's recipes are simple, practical, and unpretentious. An entire section is devoted to cooking for the sick Jesuit priests, novices, and students, but he also notes that all the recipes in the collection can be made for the sick by adjusting the seasonings and ingredients in accordance with a physician's advice. His recipe for tomatoes is one of several dealing with sauces and condiments.

Gaudentio's inclusion of tomatoes in his collection suggests that he did not consider them unhealthy. Moreover, he refers to tomatoes being "cultivated in gardens," pointing to their increasing availability. It is easy, in fact, to imagine the occasional tomato plant finding space in the market gardens and plantations of specialized plants and shrubs that extended around Italian towns, the "Mediterranean garden" described by Emilio Sereni. The tomato's transition to the open field alongside the staple crops was still in the future, however.

The tomato's reputation was changing, although it is difficult to pinpoint the time of this change and find details of how tomatoes were being eaten.

Other than Gaudentio's own unpublished work, no Italian cookbooks were published between 1694 and 1773. The only cookbooks to appear were translations of French works, testimony to the popularity of French cuisine at this time. In France, tomatoes were just beginning to make the journey north from Provence. In 1772, the Vilmorin-Andrieux seed catalog lists the tomato as a "vegetable" for the first time, whereas previously it was considered an ornamental. In the following year in Italy, the chef Vincenzo Corrado, although still careful to remove the tomato's harmful skin and seeds before cooking it, acknowledged that tomatoes were "to be enjoyed" (figure 15). Corrado accordingly included tomatoes in many of the recipes in his cookbook *Il cuoco galante, opera meccanica* (1773), all meant to accompany meat, fish, and eggs. Corrado was a Celestine Benedictine monk in Naples, and as a young man he had traveled throughout Italy from 1758 to 1760, visiting the Benedictine order's houses. His cookbooks illustrate how the cultivation and consumption of the tomato was spreading throughout Italy in the mid-eighteenth century.

Figure 15 Forty-year-old Vincenzo Corrado, dressed for work. (From Corrado, *Del cibo pitagorico ovvero erbaceo* [Naples, 1781])

In 1784, in *Del cibo pitagorico ovvero erbaceo*, his book on cooking vegetables, Corrado wrote that tomatoes were not only tasty but good for us, too. "According to the physicians," Corrado wrote, "their acidic juice aids digestion considerably, especially in their own summer season, when man's stomach is loosened and nauseous because of the great heat." Some doubts about tomatoes may have remained, however, for a 1796 agricultural treatise felt obliged to reassure its readers that the Italian climate rendered tomatoes harmless. In addition, newer varieties were of a "delicate taste and less acidic."

Corrado's book usually is cited as the first time tomatoes enter the Italian diet, but in fact, he was building on an established culinary tradition. Other sources suggest that by at least the mid-eighteenth century, tomatoes already were widely used, albeit in limited ways and apparently by only the social elites. Tomatoes also were being used in ways quite different from how they were in Mexico, which was essentially how both Latini and Gaudentio had prepared them. For example, by the late 1750s the Jesuits of the Casa Professa in Rome were eating tomatoes routinely on Fridays, meatless of course, during the month of July, in the form of *frittata con pomi d'oro*. For them, tomatoes were a welcome addition to the usual Friday lunch omelet. On other days, tomatoes might be in a *pasticciata*, encased in pastry, to accompany boiled meat or veal. In sum, tomatoes were being adapted into Italian culture, blended with local ingredients and culinary techniques; that is, they were being acculturated.

Many religious institutions, like convents, produced food for their own consumption and for others outside the convent walls, so they often established new culinary habits. For example, the Celestine nuns of Trani were fond of *brodetto al pomodoro* (tomato soup); in fact, they ate it on twenty occasions in 1751. Later, in the 1770s and 1780s, the Benedictine nuns of the Sicilian city of Catania were eating tomatoes in daintier ways: either as an antipasto called a *mortaretto*, a light pastry filled with tomatoes and herbs, or as an accompaniment to eggs and anchovies as a *terza cosa*. This "third thing" was a varied course, either a savory or a sweet, served at supper after the soup and fish courses. This Benedictine convent was rich and had the most varied and elaborate diet of all the city's religious orders, exhibiting the French influence characteristic of Sicilian aristocratic cooking of that time. Some two centuries earlier, the Italians supposedly had taught the French how to prepare food, so now the French were returning the favor, along with the notion that for those who could afford it, the enjoyment of fine food was less a capital sin than a sign of good taste.

The tomato's increasing use in Italy made an impression on foreign visitors, one of whom was Peter Collinson, a London merchant, botanist, and fellow of the Royal Society. In 1742, he wrote the following in a letter to John Custis of Williamsburg, Virginia:

> Apples of Love are very much used in Italy to putt when ripe into their brooths and soups giving it a pretty tart taste. A lady just come from Leghorn [Livorno, Tuscany's main port] says she thinks it gives an agreeable tartness and relish to them and she likes it much. They call it tamiata. I never yet tried the experiment but I think to do it. They putt in but one or two at a time, the boiling breaks them and then they are diffused through the whole.

This culinary use of the tomato reminds us of Latini's recipe of fifty years earlier, but here it was novel, so much so that tasting it was an "experiment" to Anglo-Saxons on both sides of the Atlantic. Even so, in the now proliferating gardening and horticultural literature of Britain and America, tomatoes still were listed under ornamentals and flowers, not under kitchen gardens.

Tomatoes now came in different varieties, from the red and yellow cherry type "used in medicine" to the larger, red variety with "furrowed sides," used by the Spanish, Italians, and Portuguese in soups, sauces, and salads, "to which they give and agreeable acid flavor." The words are those of Philip Miller, gardener to the London apothecaries' guild and "member of the botanic academy at Florence," in his book *The Gardener's Dictionary* (1768). The large, furrowed tomato was the original variety obtained by early botanists and was still the most common in Italy and the one represented in Italian still-life paintings of the mid-eighteenth century. The tomatillo is no longer mentioned.

The Sardinians, at all social levels, may have been the first to take the tomato seriously, perhaps because the island was a Spanish possession until 1720. The Sardinians called them *tumatas*, close to the Spanish *tomate*. An anonymous Piedmontese writer encountered the tomato for the first time while visiting Sardinia in the 1750s. The island was now a Piedmontese possession, ruled from Turin. In *Descrizione dell'isola di Sardegna*, he wrote that "this fruit is of a dark red, its shape is round like an apple, its taste sour; the Sardinians cook it with meat and eat it as a soup," and "the Piedmontese cooks [resident on the island] make exquisite sauces from it."

Curiously, a Sardinian was credited with being the first person to eat tomatoes, mixing tomato juice with beef gravy, in the "Old Northwest" of the United States (what became Illinois and Indiana). His name was Francis Vigo, and he had served in the Spanish army before establishing himself as a merchant in the 1780s. Vigo may have encountered the tomato anywhere in his travels from Sardinia through Cuba and Louisiana, all places where it was consumed.

A more reliable guide to the preparation of the tomato in Sardinia is an anonymous agricultural treatise from the mid-eighteenth century, which contains two Sardinian tomato recipes, both ways to preserve them. These recipes are a milestone in the tomato's history in Italy, as the ease with which tomatoes can be preserved and the various ways of doing it were crucial to their "success." The first recipe, "in the Spanish style," mixed sour grapes, chilies, and "less than ripe tomatoes" and covered them in vinegar. The result was a condiment to be added to sauces and stews throughout the year. The tartness of the condiment recalls the earliest Italian tomato recipes.

The second Sardinian recipe uses sun-dried tomatoes, which is perhaps the earliest reference to drying as a means of preservation. Because tomatoes were available fresh only from July to October, it was necessary to find ways to preserve them for kitchen use throughout the rest of the year. Foods had been dried as a means of preservation since ancient Greece. Using that method, the anonymous agricultural treatise advised, "Gather the ripe and round whole tomatoes before it rains, cut them down the middle and, so sliced . . . put a little bit of salt on each half, and [when they are] dried, bottle them, to be used throughout the year." By 1780, the salted, sun-dried tomatoes were being ground for use later as a seasoning. From the early nineteenth century, the method of sun-drying tomatoes is recorded also in the Neapolitan province of Principato Citeriore, as well as sun-drying eggplants, chilies, and whole melon skins, the last to be used as salt containers.

By the 1830s, but probably earlier too, enterprising peasant women in the Cagliari area were selling sun-dried tomatoes. This is an important reminder of the role of gender in agrarian change. Indeed, women frequently were responsible for the cultivation, preparation, and sale of foodstuffs, and tomatoes were becoming an important element of domestic production, if not consumption. After all, they were being sold, not necessarily eaten, in order to bring in extra income as part of the varied economic activity of the rural household. The implements required for sun-drying already were in the rural kitchen, such as the drying racks made of canes tied together and the clay cooking pots.

Another, equally common, way to preserve tomatoes for use throughout the year was reducing them into a thick paste. Before Nicolas Appert and the advent of bottled preserves, these pastes necessarily had to be quite dry. In the words of the agronomist Filippo Re, "Not only is the tomato used when it is fresh, but a *conserva* is obtained from its juice, which is thickened over the fire, then reduced to a solid consistency, and much used on choice dishes [*manicaretti*] throughout the year." This paste came to be known as a *conserva nera* (black or dark preserve). If this is hard to imagine, an American cookbook from 1847 provides a few details (although the addition of other vegetables and spices is new). A dense preserve like that in the cookbook was used as a seasoning for soups

Italian Tomato Paste

Take a peck [9.6 quarts] of tomatoes; break them and put them to boil with celery, four carrots, two onions, three table-spoonfuls of salt, six whole peppers, six cloves, and a stick of cinnamon; let them boil together (stirring all the time) until well done, and in a fit state to pass through a sieve; then boil the pulp until it becomes thick, skimming all the time. Then spread the jelly upon large plates or dishes, about half an inch thick; let it dry in the sun or oven. When quite dry, detach it from the dishes or plates, place it upon sheets of paper, and roll them up. In using the paste, dissolve it first in a little water or broth. Three inches square of paste is enough to flavor two quarts of soup. Care should be taken to keep the rolls of paste where they will be preserved as much as possible from moisture.

Sarah Rutledge, *The Carolina Housewife* (Charleston, S.C.: Babcock, 1847), 105, reprinted in Andrew Smith, *The Tomato in America: Early History, Culture, and Cookery* (Urbana: University of Illinois Press, 2001), 177.

and meats and in sauces. It appears in the kitchen account books of the Saluzzo family, dukes of Corigliano (Calabria), for the years 1789 to 1791, as *salsa* or *conserva di pomodoro*. It is listed alongside other spices and flavorings like cinnamon, nutmeg, and black pepper, which is an indication of how it was perceived.

Corrado, the Benedictine monk and cookbook author, helps fill in the details of these references to the use of tomatoes. For example, his recipe for *salsa di pomodoro* uses fresh tomatoes, not dried, which then are cooked. When the *salsa* is to be served over mutton, mutton gravy is added, and when the *salsa di pomodoro* is to accompany scorpion fish, it is prepared with butter, garlic, and bay leaf. These two suggested uses correspond to what Corrado elsewhere calls his two versions of tomato condiment: with and without meat. The latter version is intended for the many meat-less or "lean" days of the Catholic calendar. In addition, Corrado recom-

A Selection of Vincenzo Corrado's Tomato Recipes

Pomidori alla Napolitana (tomatoes, Neapolitan style). After cleaning the tomatoes of their skins and cutting them in half, you will remove their seeds and place them on a sheet of paper greased with oil in a baking tray. You will fill the tomato halves with anchovies, parsley, oregano, and garlic, all finely chopped and seasoned with salt and pepper. Having covered the tomatoes with bread crumbs, you will bake them in the oven, and serve them.

•••••

Zuppa alli pomodoro (tomato soup). Cook a quantity of tomatoes in a beef broth with a small bunch of aromatic herbs. After clarifying this broth, add toasted bread crusts, and serve with a tomato *coli* seasoned with basil, thyme, and parsley, on top.

•••••

Salsa di pomodoro (tomato sauce). After cleaning the tomatoes of their skins and seeds, you will chop them together with garlic cloves, red chilies, pennyroyal, and rue. After straining this mixture with the addition of oil and seasoning it with spices, boil it with vinegar and mutton sauce, and serve it warm over mutton.

Vincenzo Corrado, *Il cuoco galante, opera meccanica* (1773; Naples: Raimondi, 1786; facsimile ed., Sala Bolognese: Arnaldo Forni, 1990), 140, 158, 189.

mends his *coli di pomodoro*—from the French *coulis*, referring to a condensed sauce—as an accompaniment to a wide variety of foods: veal, sliced veal head, stuffed small chicken, turtledove, roast sturgeon, trout, sliced crayfish, poached eggs, and squash.

It is interesting that in many respects, Corrado's recipe for *zucche lunghe alla parmegiana*—he did not approve of eggplants—otherwise resembles the modern one for "eggplant Parmesan." The squash is cut in round slices, salted to remove the liquid, fried (in pork fat, not olive oil), layered between butter and Parmesan cheese, and covered in a sauce made of eggs and butter before being baked. Later, tomato sauce found its way into the dish, and the squash was replaced by eggplant.

Corrado also profited from the new popularity of vegetables with his cookbook *Del cibo pitagorico ovvero erbaceo* (1781). As a seasoning for other vegetables, he recommends his *purè di pomodoro*. This is prepared by quartering the tomatoes; sautéing them in fat or oil with garlic, parsley, radish, bay leaf, and celery; and adding some broth and, when this boils, bread crusts, all of which is strained through a sieve. He describes this ancestor of the modern *passata* as a "tasty condiment, almost necessary to impart greater flavor to numerous vegetable dishes." As this comment suggests, Corrado's book was not meant as a guide to an austere and meatless Pythagorean diet in line with Antonio Cocchi's recommendations, as discussed in chapter 2. Rather, it was intended to provide a way to make vegetables tastier. Corrado also suggests using tomatoes as a separate dish, as in his earlier recipe for *pomidori alla Napolitana*. For this, the tomatoes first had to be peeled, by placing them on embers or in boiling water, and then the seeds had to be removed. The tomatoes then were halved, stuffed with various fillings, and either baked or fried. He recommended round, yellow tomatoes (which he called saffron colored).

The tomato still had only a regional presence, limited to only a few areas. Nonetheless, in the tomato's introduction to and early uses in Italy, the Spanish influence is clear. Most of the preceding references to tomatoes come from the south of the Italian peninsula, as well as the islands of Sardinia and Sicily, all formerly Spanish dominions. Tomatoes also could be found in Tuscany, where our story began, and cookbooks from Rome and Macerata suggest a presence in other parts of central Italy, like Lazio and the Marche.

The tomato's use in northern Italy was more sporadic. Only in Liguria was it becoming an important agricultural product, part of the already thriving trade in fresh vegetables and fruit referred to in chapter 2. Household accounts show that in 1765 the Doria di Montaldeo, an aristocratic Genoese family, purchased tomatoes between July and September and then again in December (perhaps dried or as *conserva*). They were listed as *tomate*, suggesting either an abiding Spanish influence or a fashionable French one.

The French influence would seem to account for the introduction of the tomato in what was to become one of Italy's main producing areas: the region around Parma. Jean-Gabriel Leblanc, chef to Napoleon's second wife, Marie Louise of Habsburg-Lorraine, used *pommes d'amour* widely in sauces. When Marie Louise refused to follow her husband into exile in 1814, the Treaty of Fontainebleau made her the duchess of Parma, Piacenza, and Guastalla, and she brought Leblanc with her when she arrived in Parma two years later (becoming in time the much-loved Maria Luigia). The tomato came with them. Leblanc's successor, Vincenzo Agnoletti, recommended, in a work published in 1832, using preserved tomatoes either dried in loaves (*mattoncini*) or as thickened sauce (*passata*), in soups or with fish and meat. Tomatoes were grown at the estate in Colorno, outside Parma, which supplied the palace with much of its produce. In August, a palace cook and two assistants would turn the tomatoes into paste. From the account books of Maria Luigia's household, we know that in 1844, three years before her death, they made eighty jars of *conserva*, weighing 86 pounds.

By this time, tomatoes had spread beyond the Colorno estate and were being grown in the area's rich farmlands.

As we enter the nineteenth century, we find more evidence of the tomato's being eaten throughout society. In the words of a Tuscan botanist, Ottaviano Targioni Tozzetti, writing in 1813, tomatoes were "cultivated in all market and kitchen gardens," and their use was "very common." Luigi Bicchierai, called "Pennino," ran an inn located at a crossing along the Arno River, at Ponte a Signa, not far from Florence. We do not know whether he was a typical innkeeper, for he had learned how to cook at a local monastery from two Neapolitan friars before taking over the family inn in 1812. Unlike most innkeepers, Bicchierai also found time to

Sugo della miseria (Poor Man's Meat Sauce)

This sauce isn't holy, but where it falls, it does miracles.

When you have made a nice selection of boiled meats, there is always some left over, and reheating it becomes a bit tiresome, likewise with meatballs.

So I undertook to create a sauce that would be pleasing. Thus, you take four good potatoes and cube them, chop an onion, three celery sticks, parsley, garlic, one leek, bay leaf, sage, cloves. Put everything in a pan to blend, with two glasses of white wine and a little broth.

Just before it is ready, add the sliced boiled or minced meat and the tomatoes, mix well, and use this in the same way as any other ragout; no one will realize it and will enjoy it all very much.

Luigi Bicchierai, *Pennino l'oste*, ed. Franco Tozzi (Signa: Masso delle Fate, 1996), 15.

keep a commonplace book, a sometimes witty mixture of accounts, diary entries, recipes, and sonnets. Here he suggested using tomatoes in his poor-man's meat sauce. In the best traditions of Tuscan frugality, the sauce uses leftover meat, along with potatoes, onions, herbs, and tomatoes. No one would be able to tell it from an authentic meat ragout, Bicchierai claimed.

Tomatoes had now become so common that people were throwing them away, or at least were throwing them. In Italy, tomatoes were the missile of choice to show disapproval of public performers, and the activity came to be known as a *pomodorata*. In July 1838, the Roman poet Giuseppe Belli composed a few poems for an upcoming recital at the city's Accademia Tiberina. Unsure how the poems would be received, Belli wrote to a friend, "God save us from the tomatoes" (*Dio ci salvi dai pomodoro*).

For the peasants of southern Italy, tomatoes were all that they had to eat during the dog days of summer. In fact, vegetables formed the basis of the southern peasants' diet, as we saw in chapter 2. The problem was that they "abounded" from late autumn through spring but, in the words of the *Statistica murattiana* of 1811, were "rare during the summer because of the lack of water." At that time, only onions, garlic, chicory, chilies, eggplants, and tomatoes were available.

Growing tomatoes was, and still is, labor intensive. The soil had to be dug deeply, and the plants had to be well fertilized, watered, supported, and pruned as they grew (although in the south, the trailing varieties were later favored). Harvesting the fruit required the contribution of the whole family. But to compensate, the tomatoes matured quickly and produced abundant fruit. Along with chilies, tomatoes offered a bit of chromatic variety, which was welcome in a peasant diet monotonously colored a brownish green. Moreover, the color red had associations of wine and meat, both precious.

Tomatoes also were cheap. According to the *Statistica*, tomatoes sold for 3 *grana* a *rotolo* (a *grano* was a small copper coin, and a *rotolo* was equal to 2 pounds). For around 40 *grana* a day, a family of five could subsist on the common mixture of bread, legumes, and vegetables, plus a glass or two of wine. At this time, vegetables might cost from 2 to 8 *grana* per *rotolo*, so tomatoes were relatively inexpensive.

For all these reasons, almost everywhere in southern Italy, tomatoes formed part of the peasants' "ordinary food," part of the "subsistence of the population," according to the *Statistica*. Tomatoes, as well as chilies and garlic, also were used as "both a food and a seasoning" by the peasants of Calabria Citeriore.

What sorts of tomatoes were cultivated? The *Statistica* tells us only that tomatoes were grown in many areas of southern Italy. For instance, they were cultivated in Terra di Lavoro, outside Naples, "with their varieties in shape." An agricultural treatise published in the same year, Filippo Re's *L'ortolano dirozzato*, fills in the details. The varieties were *schiacciato* (squashed or flattened), *globoso* (spherical), and *peretto*. The last is, in fact, the first mention of a new variety, the pear-shaped tomato, which was destined for great things later in the century.

In Basilicata, peasants took their tomatoes to market, to provide much-needed income. In Potenza, the sale of tomatoes and other vegetables by peasants from the surrounding countryside formed a retail trade. Peasants also supplied the towns of the district of Melfi, such was the "abundance" of produce. This trade was necessarily localized, given the poor means of transport, and the peasants around Naples stood to benefit the most. As a foretaste of things to come later in the century, in the Terra di Lavoro, tomatoes (along with chilies and cucumbers) were said to "offer great resources for the domestic economy of all families during the summer months."

Tomato cultivation remained a cottage industry, outside the concerns of state governments. In the years when most of the Italian states were establishing small-scale "experimental farms" to encourage the production of cash crops, tomatoes are not mentioned at all. Instead, the farms focused their research and trials on plants of presumed economic importance, from grains and feed crops to plants used in the textile and dyeing industries. In the same way, local seed banks (*monti frumentari*), which provided peasant farmers with seeds for cultivation, concentrated on the staples, primarily wheat, but increasingly also legumes, potatoes, and maize.

The cultivation and consumption of tomatoes increased markedly during the nineteenth century. In Naples, this was paradoxically based on a decline in overall food consumption. There is evidence that the already poor diet of many southern Italians actually worsened after the unification of Italy in 1861. In Naples, now reduced to the status of a provincial capital, the per capita consumption of basic foodstuffs like meat and cereals fell markedly. Consequently, the urban poor compensated by eating still more fruits and vegetables. Producers were able to meet the rising demand by aiming for quantity over quality, thus cultivating cheaper, inferior varieties. The number of fruit and vegetable sellers in the city increased almost threefold between 1845 and 1881, at a time when the overall population rose only slightly. As a result, the diet of the poor was almost exclusively vegetarian.

This included tomatoes, even eaten raw! Culinary custom advised that tomatoes be cooked not once but twice. The first cooking was to remove the skin and seeds, and the second was to cook the tomatoes into a sauce or as a dish on their own. But two Neapolitan physicians, Achille Spatuzzi and Luigi Somma, writing about the health of the city's vast proletariat, noted their "very great use" of tomatoes: "In the summer months, while [the tomatoes] are still unripe, they are eaten raw in a salad with onion, oregano, and so on." This was part of a widespread diet of cheap unripe or rotten vegetables and fruits and resulted in the predominance of gastrointestinal complaints and indigestion during this time of the year. The doctors were right to be worried about the state of the city's health. Thirty years earlier, Naples had suffered a severe epidemic of cholera, which had hit the poorest and most densely crowded areas the hardest.

The two doctors were not against tomatoes per se. On the contrary, "aside from the unripe tomatoes," they judged them "valuable" and their use "praiseworthy," especially as a condiment. But there was a clear class division. The city's poor were eating tomatoes raw, either unripe or rotten, in salads, and setting aside the good tomatoes to be cooked in sauces or reduced into a paste for winter use.

Finally, and on a more positive note, the Neapolitan doctors offer us invaluable evidence of two local uses of tomatoes for which Italy would soon become world famous. They refer to the poor's subsistence on something called *pizza* (and they italicized the word, for the particularly Neapolitan form they had in mind had not yet entered the Italian or any other language). They explained for the benefit of their readers that the *pizza* was "seasoned on the top with an abundance of oil or pork fat, with cheese, oregano, garlic, parsley, mint leaves, with tomato especially in summer, and finally sometimes even with small fresh fish." Tomato was not a yet a basic element of pizza, but only one possibility among several.

For those who could afford it, tomatoes appeared in another new guise: "They form the customary seasoning for macaroni, and not a day goes by when they don't appear on the tables of the middle class." This is perhaps the earliest reference to pasta as a staple food, accompanied by a tomato sauce.

4

Pasta al Pomodoro

At the beginning of 1849, the Tuscan innkeeper Luigi Bicchierai, "Pennino" to his friends, noted the following in his commonplace book:

> With all these rebellious uprisings and the desire to create Italy, I, who am an innkeeper and know little of such things, have thought about Italy so divided but which everyone wants, including me; and I imagine it like a lovely pot of boiled meats: trotter, tongue, various cuts, and seasonings! And so, if Italy is a *bollito*, its flag should be the condiment for it, in other words, the "Tricolor Sauce."

Bicchierai goes on to provide the recipe for this patriotic sauce of his own invention, in the colors of the flag—green, white, and red—proposed for a united Italy. It consists of three separate sauces. The *salsa verde* is a mixture of finely chopped capers, anchovies, hard-boiled eggs, basil, parsley, onion, and garlic, moistened with olive oil and wine vinegar. It was (and is) found in many regions of Italy as an accompaniment to mixed boiled meats (*bollito misto*). The *salsa bianca* is a simple béchamel sauce, but it is Bicchierai's *salsa rossa*, you will have guessed, that interests us most.

    Bicchierai's "tricolor sauce" is an appropriate way to start this chapter on the nineteenth century. In Italy, the nineteenth century was shaped by the movement to unify the different states of the peninsula and islands

---

Salsa rossa (Red sauce)

The last color of the [Italian] flag, the *salsa rossa*: Put in a pan seven or eight large tomatoes, chopped; one-quarter onion; two basil leaves; a stick of celery, finely chopped; and a little parsley. When the tomatoes have lost much of their liquid, strain everything through a sieve; pour this *passata* into a pan; add a spoonful of [olive] oil, a pinch of salt, and a small pinch of pepper; and bring it to a boil, occasionally stirring it. The lovely red sauce is ready. If you want it sweet-and-sour, all you have to do is stir in a large spoonful of [wine] vinegar and a small spoonful of sugar before removing the pan from the fire; taste; and add more vinegar or sugar if necessary.

Luigi Bicchierai, *Pennino l'oste*, ed. Franco Tozzi (Signa: Masso delle Fate, 1996), 105–6.

---

into a single great nation, the Risorgimento. By 1861, the process was largely completed, although parts of the Papal States, and Rome itself, were not annexed until 1870. More important, this was the century when pasta and tomato sauce were fatefully combined to create *pasta al pomodoro*.

On July 26, 1860, commenting on the project to unite Italy, one of its proponents, Camillo Benso, count of Cavour, wrote in a kind of ironic code and in French that while "les oranges [the Sicilians] . . . sont déjà sur notre table," "les macaronis [the Neapolitans] ne sont encore cuits." The Sicilians, in other words, were ready for Giuseppe Garibaldi's expedition and eventual annexation to a united Italy, under King Victor Emanuel of Savoy. But the Neapolitans, "the macaronis," were not. Naples already was synonymous with pasta, especially dried pasta made from durum semolina wheat, known simply as "Naples pasta" (*pasta di Napoli*) or "Neapolitan macaroni" (*maccheroni napolitani*). The House of Savoy may have conquered Naples in the nineteenth century, but the consumption of Neapolitan-style pasta would conquer all of Italy by the late twentieth century.

To understand the significance of the encounter between pasta and tomato—the how, when, where, and even why of how this classic combination originated—we must look briefly at the history of pasta in Italy. We do not have to bother with the debates about who invented or discovered pasta. In fact, pasta, in the sense of any product made from a dough made from mixing wheat flour (hard or soft, with the bran removed) and water (or eggs), has been found at many different times and in many places, from the Mediterranean to China. Suffice it to say that various forms of pasta were well known in Italy by the time the tomato first appeared in the mid-sixteenth century: lasagne, vermicelli, *maccheroni*, to name only the main types.

This pasta was very different from what we have today. First, it was eaten soft. Since the Middle Ages, pasta had been cooked for long periods—half an hour, an hour, even two hours—generally in broth, until it was soft and "melting" (*fondente*). This was the standard way of cooking both fresh and dried pastas, which conformed to then current notions of taste and health. Second, because pasta was served moist, it usually was accompanied by dry condiments, and it was seasoned in different ways, "whether with oil or walnuts or almonds, or with milk or cheese, or with pepper or other spices," in the words of Costanzo Felici. Only pork fat is absent from Felici's list. Third, pasta was not yet eaten as a dish in its own right—as a first course, say. Instead, it was served as a side dish (for example, to accompany meat) or as a dessert.

Travelers visiting in Italy were often struck by the place of pasta in the Italian diet. John Ray, one of the first Englishmen to observe personally the Italian consumption of the tomato, described the use of pasta in some detail. Note that he uses the term *pasta* in the generic way we use it today:

> *Paste* made into strings like pack-thread or thongs of white leather (which if greater they call *Macaroni*, if lesser *Vermicelli*), they cut in pieces and put in their pots as we do oat-meal to make their *menestra* or broth of, much esteemed by the common people. These, boiled and oiled with a little cheese scraped upon them, they eat as we do

buttered wheat or rice. The making of these is a trade and mystery [organized craft]; and in every great town you shall see several shops of them.

By the time of Ray's visit, pasta was being made throughout the Italian peninsula and islands, usually in the vicinity of flour mills and bread bakers. Pasta makers had been granted their own guilds, independent of the bakers' guilds, in cities like Naples, Palermo, Genoa, Savona, and Rome. Strangely, Sardinia is absent from this list, even though the pasta of Cagliari already had a favorable reputation. Otherwise, the establishment of guilds reflects the regionality of pasta making, and all these areas became famous for their production as the industry continued to grow.

The two best-known areas of production were the Ligurian coast from Genoa southward and the Bay of Naples and Amalfi coast. Entire towns were devoted to making pasta. The Ligurian town of Port Maurizio (today's Imperia) had forty pasta manufactories by the beginning of the nineteenth century, supporting two hundred families. In the vicinity of Naples, places like Torre Annunziata, Gragnano, and Castellammare, already the location of numerous mills, effectively became one-industry towns. By 1633, the Naples region was already exporting more than 140 *cantaia* (some 30,000 pounds) of pasta.

The specializations of the two regions, Naples and Genoa, were quite different from each other, at least in the minds of contemporaries. When François Chapusot, chef to the English ambassador at the Savoyard court in Turin, prepared macaroni, he used only that imported from Naples. "It is generally the best, both for the chosen quality of wheat that is employed, and also because it can be boiled without breaking up into pieces," Chapusot wrote. But when choosing finer and smaller pasta shapes for soups, he preferred pasta from Genoa.

By the seventeenth century, the teeming metropolis of Naples already was identified with pasta production and consumption, and the city's guild of vermicelli makers (Vermicellari) dates from at least 1546. The vermicelli makers joined the new guild of macaroni makers (Maccaronari) in 1699, a trend exemplifying how the term *maccherone* came to stand for all forms of pasta. The Frenchman Jérôme de Lalande, traveling in Italy in the 1760s, differentiated more than thirty pasta shapes being made in the Naples region, as well as listing the various presses and cutters used to make them.

Pulcinella, a comic theater character associated with Naples, was constantly depicted eating pasta (figure 16). In a 1632 commedia dell'arte play,

Figure 16 Giuseppe Garibaldi's small force routs the army of Bourbon Naples—depicted as a mass of feckless, macaroni-touting Pulcinellas—in Cabrion's [Nicola Sanesi] "Attualità: L'ultimo Atto di un Dramma!" (Current Affairs: The Play's Final Act), published in the satirical newspaper *Il Lampione*, September 4, 1860, 3.

Silvio Fiorillo has Pulcinella condemn the city's overlords (portrayed by a Spanish soldier) with the words: "Ah, Spaniard, enemy of macaroni," as if to say, the enemy of the Neapolitans themselves. When the ever famished Pulcinella dreams at night, he tells Clarice, with whom he is in love,

> I dream—oh, love and what it does!—of a big dish of macaroni with meatballs on top. I reach out my hand, grab hold of the macaroni and meatballs, I season them, I mix them up, and when I go to put them in my mouth, I wake up all in a huff . . . with passion in my heart I start weeping like a child.

A French visitor to the city recorded a play in which Pulcinella, having become king, was told that he would have to give up his macaroni, as it was considered too plebian. To this Pulcinella replied, in Neapolitan, "Mo mo me sprencepo" ("Then I resign, effective immediately," or words to that effect).

In Naples, freshly made pasta was laid out on cane racks or large cloths to dry in the sun. If this seems like an intriguing parallel to drying tomatoes, remember that drying and salting were the main methods of preserving food. Drying pasta was actually a complex process, with three drying phases of different duration according to the shape of pasta and the weather. The method was perfectly suited to Naples's temperate climate. It was said that pasta should be made when the warm, humid southerly wind blew and be dried when the drier and cooler northerly wind blew (*i maccheroni si fanno col sirocco e si asciugano con la tramontana*). As a result, the process was almost impossible to duplicate elsewhere in Europe.

By the 1790s, the elites of Italy were eating *maccheroni napoletani* on special days, some as part of the "first course," which itself was a recent innovation, since previously the dishes had been served at the same time. At the Milano-Franco household in Polistena (Calabria), "upstairs" might have a dish of *marmitta di maccaroni*, plus various other dishes; and "downstairs" had simple macaroni, soup, and boiled meat. The servants were not eating pasta as a separate course yet, but they certainly were eating it.

According to the *Statistica murattiana* of 1811, the better-off of Potenza (Basilicata) ate pasta brought from the Amalfi coast, south of Naples. The poor, by contrast, either made their own or bought it at half the price from the local *maccaronari*, "who make it badly." For southern Italian peasants, pasta was a holiday luxury, not an everyday staple. Nonetheless, it had become the "symbol of joy and abundance," according to the compiler of the Molise submission for the *Statistica*. The different seasonings they used reflect an almost medieval mixture of sweet and savory: "Rarely do the people season it with cheese, but with oil, pork fat, or else with vinegar or cooked grape must."

For more affluent people, pasta was fast becoming an everyday food, beginning in the grain-producing regions of the south. Religious institutions, like aristocratic households, would sign annual contracts with local pasta makers, who supplied them regularly for a fixed price. In 1817, the ten Franciscan friars at the monastery of San Matteo, in San Marco in Lamis (in the grain-rich plain around Foggia, Puglia), managed to eat their way through 392 pounds of *maccheroni*, 353 pounds of *maccheroncini*, 71 pounds of *pasta fina*, 11 pounds of *vermicini*, and 9 pounds of *punta d'ago*. Although these are different forms for different uses, these figures work out to roughly a dish of pasta (3.7 ounces) a person per day. Female religious ate large quantities too, but in more elaborate forms like *gnocchetti, taglioni, rafaioli,* and egg pasta. According to the form of pasta, it was eaten with vegetables, in broth, seasoned with *lardo* (hard bacon fat) or cheese, or, on special occasions, with a tomato-less veal *ragù*. Nuns tended to make their own, and some convents became well known for their handmade pasta.

Pasta triumphed in nineteenth-century Naples. At the time of Italy's unification, Neapolitans were great consumers of pasta, locally made and of excellent quality, according to Achille Spatuzzi and Luigi Somma, the two Neapolitan doctors quoted in chapter 3. Even the poor could afford to eat pasta, although they had to make do with the so-called *minuzzaglie,* the damaged bin-ends of different pasta shapes and sizes, at a time when pasta was still sold loose.

By the mid-nineteenth century, pasta was eaten in taverns served with a meat or tomato sauce and topped with grated hard cheese. According to Carlo Tito Dal Bono's "Le taverne" (1858), a description of Naples's taverns, the macaroni eater "enjoys a meat sauce, but the tart tomato is perhaps equally pleasing, and where one or the other is absent, he sticks to the purest simplicity, of cheese alone." Pasta was eaten in the backs of shops by shopkeepers supplied with ready-made portions by the city's macaroni makers and sellers. In 1845, Naples had 280 *maccaronari*.

It was as street food that pasta came into its own. While visiting the city in 1787, Goethe commented: "The macaroni . . . can be bought everywhere and in all the shops for very little money. As a rule, it is simply cooked in water and seasoned with grated cheese." Pasta was consumed primarily in the open, bought cheaply from stalls in the streets and markets equipped with boiling cauldrons, where it was quickly seasoned and consumed, using the hands, as depicted in the inexpensive "genre" prints of the time and (later) picture postcards for tourist consumption (figure 17). The Neapolitan journalist Matilde Serao described the scene in *Il ventre di Napoli* (1884):

> As soon as he has two cents, the Neapolitan pleb buys a dish of macaroni, cooked and seasoned. All the streets of the city's popular quarters have one of these taverns that set up their cauldrons outside, where macaroni are always on the boil, with pots of simmering tomato sauce and mountains of grated cheese. . . . This setup is very picturesque, and a few painters have painted it . . . and in the collections of Neapolitan photographs that the English buy, alongside the lay nun, the petty thief, the flea-ridden family, there is always the macaroni seller's table.

This kind of "street food" approach to the preparation and consumption of pasta was probably responsible for another important

Figure 17 Gaetano Dura, *Mangia Maccaroni* (*Macaroni Eaters*) (*above*) and *Famiglia di Marinari* (*Family of Sailors*) (*below*). From a series of lithographs depicting popular Neapolitan scenes, printed by Federico Gatti, Naples, ca. 1840. (© Trustees of the British Museum)

development: an appreciation of eating the pasta still slightly hard, not completely cooked. This was a novelty. The new Neapolitan practice particularly suited the local pasta, which was made from durum wheat and dried, as it allowed it to keep its elasticity and slightly chewy consistency. Writing in *Cucina teorico-pratica* (1837), his guide to Neapolitan "home cooking," Ippolito Cavalcanti, duke of Buonvicino, insisted that the macaroni be cooked only until *vierd vierd*, a Neapolitan expression meaning "slightly hard" (literally, "very green"). After Italy was unified, this practice spread to the rest of Italy. In a cookbook to which we shall return later in this chapter, Pellegrino Artusi advises that "as for the macaroni themselves, the Neapolitans recommend that they should be boiled in a large pot, with lots of water, and not cooked too long." The pasta shapes should be cooked "fairly hard" (*durette*). Paradoxically, and defying the dietary advice of previous centuries, this actually makes pasta easier to digest. As Artusi explained, the slight hardness requires that it be chewed before being swallowed. But he does not call this consistency *al dente*, that untranslatable expression referring to this brief cooking. That expression did not become common in Italy until after World War I.

The gradual industrialization of pasta making in the second half of the nineteenth century helped turn it into a staple. How does pasta making compare with tomato production? Whereas pasta making was an Italian—especially a Neapolitan—success story, food preservation remained a mainly domestic or cottage industry well into the twentieth century.

The simplest method of preserving tomatoes was hanging the entire plant before it had ripened. The tomatoes would gain in sweetness over the following months and could be used "fresh" during winter. A manual on vegetable cultivation published by the horticulturalist brothers Marcellino and Giuseppe Roda, *Manuale dell' ortolano contenente la coltivazione ordinaria e forzata delle piante d'ortaggio* (1868), noted that people saved healthy, but still partially green, tomatoes in the autumn, which soon became a common wintertime decoration for rafters and walls. In the words of

Marcellin Pellet, the French consul general in Naples, writing in 1894: "The diet [of the inhabitants] is essentially vegetable. On the walls of houses, on either sides of the windows, one sees hanging long bunches of tomatoes, onions, poles of prickly pears [a kind of cactus with edible fruits]." Particularly suited to this practice were the small, firm, thick-skinned tomato southern varieties like 'Principe Borghese', which dates from this period.

Next in the hierarchy of complexity was halving, salting, and drying the tomatoes, a method first practiced in Sardinia. But more commonly, the fresh tomatoes were transformed. According to the Roda brothers, they were turned into a *pasta* (paste), "like a soft polenta," by extended simmering and then straining. This paste could then be dried on boards in the sun for three or four days and then rolled flat, to produce a more concentrated *conserva nera* (the older method we encountered in chapter 3). Or the paste could be bottled, without further drying. The bottles would be sealed and boiled in a bain-marie, according to the Appert method. The Rodas favored the second option, "since the paste maintains its lovely original color in the bottles, with a flavor not dissimilar from the fresh fruits." This is today's *passata*, in the sense of having been "passed" through a strainer, although early-twentieth-century manufacturers some-times called it *salsa*.

For much of the nineteenth century, preserving tomatoes remained a small-scale, local activity that was closely linked to domestic agricultural production. At this time, the agricultural sector dominated the young nation's economy, employing seven out of ten people at the time of Italy's unification in 1861 and even still employing half of all workingmen and -women on the eve of World War II. Dietary practices were dictated by necessity and the need to economize. At the end of the nineteenth century, food accounted for around 75 percent of rural and urban working families' living expenses, a figure that did not dip below 50 percent until the early 1960s.

Preserving tomatoes was important in symbolic terms, too. In Giovanni Verga's realist novel *I malavoglia* (1881), Rosolina's *conserva dei pomidoro* crops up throughout the novel, like a refrain echoing the decline of the family at the heart of the book. Indeed, food imagery and references permeate Verga's novel. Although the foods are basic, the language is rich in sayings, proverbs, and metaphors and is accompanied by Verga's use of repetition and leitmotiv, capturing the flow of life and the passage of time.

Near the beginning of *I malavoglia*, the unmarried Rosolina boasts to her brother, a priest, of her skills as a housekeeper. This includes "the tomato purée to make; that she knew a special way of making it that kept it fresh right through the winter" (chapter 4). Halfway through the book, some women are having an animated discussion "while Rosolina was cooking her tomato purée, her sleeves rolled up" (chapter 10). Later, thinking that a certain Don Michele is interested in her and that he will walk by her house just to catch a glimpse of her, Rosolina "was perpetually busy on her balcony with her tomato purée or bowls of chilies, to show what she was capable of" (chapter 13). But Don Michele is not interested in her, another woman says: "Donna Rosolina can make eyes at him from her balcony until she takes root among her tomatoes, for all the good it will do" (chapter 13).

At the end of the book, with the Malavoglia family utterly ruined, Rosolina is in a bad way as well:

> And Donna Rosolina had lost weight through bad temper, especially after Don Michele had gone away and all his misdoings had come to light. Now she did nothing but go about in search of masses and confessors, here and there, as far as Ognino and Aci Castello, and she neglected her tomato purée and tuna in oil, to dedicate herself to God. (chapter 15)

Verga singles out Rosolina's isolation from the start of the book when she is making *conserva* on her own. By the time Verga was writing, though, its preparation was very much a group ritual. The French archaeologist and antiquarian François Lenormant, traveling through Catanzaro

(Calabria), was struck by the "enormous pyramids" of tomatoes at the town's market. Captivated, Lenormant had this to say about the domestic preparation of tomato preserve:

> We are, in effect, in the season in which, in every Calabrian house, tomato preserve is made for use during the rest of the year. It is a solemn occasion in the popular life of these lands, a kind of festive celebration, an excuse for get-togethers and gatherings. . . . Neighbors, and especially the neighborhood women, get together in different houses one after the other for the making of *conserva di pomi d'or*, a procedure that culminates with a large meal; and they gossip as much as they can while crushing and cooking the tomatoes. It is here that for several months the locale's chronicle of scandal is identified and commented on; it is here that those old rustic songs, which are today so avidly collected by scholars keen on folklore, are repeated from generation to generation.

A new ritual, based around a new foodstuff, thus took on an old form.

Sicilian families soon carried across the Atlantic the process for making what they called *astrattu* (extract) (figure 18). This process was described by Carlo Levi, a Turinese sentenced to "internal exile" in Basilicata in 1935/1936. Even though it may be unappetizing, Levi conveys the scene as if in a still-life painting, which was no coincidence, given that Levi was also an artist:

> Out on the street, in wide-rimmed tables below the black pennants decorating the front doors, blood-red liquid masses of tomato conserve lay drying. Swarms of flies walked without wetting their feet over the portions already solidified, in numbers as vast as those of the children of Israel, while other swarms plunged into the watery Red Sea, where they were caught and drowned like Pharaoh's armies as they hotly pursued their prey.

Figure 18 The drying phase in the preparation of traditional tomato paste (*astrattu*) in Canicattì, Sicily. (Photograph by Diego Sgammeglia, courtesy of Alfonso Messina)

In the early twentieth century, the inhabitants of the malaria-ridden countryside south of Rome even resorted to removing and using as strainers the window screens that the health authorities had installed to keep out the mosquitoes, a sign of the peasants' mistrust of the medical authorities, to say nothing of the importance of tomato *conserva* to their diets.

The making of tomato *conserva*, in large vats left out in the sun and stirred from time to time to mix the slowly thickening tomatoes (and to chase away the flies), was also a feature of urban balconies. In *Racconti napoletani* (1889), Serao describes the "acrid smell of tomato *conserva*" coming from the balconies of the "vicoletto delle Gratelle" in Naples's dense network of narrow, ancient streets. The sun did the work of cooking over the fire in condensing the tomato sauce into a purée, which might then be formed into loaves or bottled for storage. Indeed, tomatoes were preserved in the form of paste "in great abundance" in the city and, from at least the 1860s, widely consumed, according to the Neapolitan doctors Spatuzzi and Somma.

Dried tomato conserve, paste, concentrate, or extract was not made only in the south. In 1849, Heirom Aime, the American vice-consul, found it in the Ligurian port of La Spezia, available in a form resembling "a large Bologna sausage." He saw a great economic potential in its manufacture and sale in the United States. In Genoa, the tomato conserve, "sufficiently stiff to stand in conical forms, as loaf sugar," was sold in food shops, brought in from the surrounding countryside by the wagonload. Less than forty years later and despite the absence of a long-standing craft tradition behind it, the preparation of tomato *conserva* did become a "profitable agrarian industry" in Liguria. The Jacini Parliamentary Inquiry of 1878, on the state of Italian agriculture, found that large quantities of the tomatoes grown around the town of Albenga—ancestors of today's large, segmented oxheart variety—were sold outside the region. As early as May, fresh tomatoes found their way by rail to the markets of Piedmont and beyond, and tomatoes later in the season were turned into paste. In fact, eight hundred to one thousand, 132-pound barrels of tomato paste were exported each year, mostly to the United States.

From its origins as a cottage industry, the preparation of *conserva dell'estratto di pomodoro* on a more industrial scale became concentrated (sorry!) where tomato cultivation was most extensive, along the Ligurian coast, in Emilia Romagna (around Piacenza and Parma), and in Campania (around Salerno and Naples).

Drying was still, however, the most common method of preserving tomatoes. Tomatoes were usually dried into a paste in the countryside where they were harvested. Thus it was unfortunate that they were poorly presented in the markets, opening them to competition from Portugal, Greece, and Turkey. This was an especially sad outcome for southern Italy and Sicily, where consumption was highest. Only two artisans from the Tuscan town of Castelfiorentino succeeded in making a product that was pleasing to the judges. This consisted of tomatoes "condensed into thin, translucent sheets, and of pleasant color and agreeable taste." Ten years later, the mayor of Miglianico (Abruzzo) presented his town's tomato *conserva* to a national parliamentary inquiry on industrial production. Before this, all its *conserva* had been consumed locally, the mayor said, but he hoped to organize its export. In 1879, Andrea Vivenza, an Abruzzese agronomist-physician, even wrote a treatise on the preparation of tomato preserve.

Other preserving methods were suggested. A successful Neapolitan cookbook, *La cucina casereccia* (1816), recommends preserving tomatoes by putting them in jars and covering them with brine, but this method never really caught on. Even though the acidity of tomatoes would have made them easier and safer to preserve than many other fruits and vegetables, it was another hundred years before canning whole tomatoes became commonplace.

Initially, the food-preserving industry did not consider canning tomatoes to be worthwhile or even technically possible. Meat, however, was canned, primarily for use by the military, as were more expensive vegetables, like peas, which had a short growing season. Preserving vegetables and fruit on an industrial scale then began in earnest in the mid-nineteenth century in Britain, the United States, and Germany. Tin cans replaced glass bottles and jars, as they were cheaper to produce and protected their contents from the light. At this time, Italy did not have much of a food industry. Preserved food products were limited to the cured meats of the center and north and the dried fruits of the south, but these were more small-scale rural domestic products and urban artisan manufactures than an industry as such. The Cirio Company, which later came to dominate the Italian production of preserved tomatoes, was still involved mainly in the export of fresh produce, using the newly laid railways. Cirio's forays into food preservation were limited to canning peas, asparagus, artichokes, peaches, and pears, for which there was a great demand. Nevertheless, the company was considered important enough for a government decree in 1885 allowing Cirio train cars to be painted with the red, white, and green of the national flag.

Tomatoes were first canned in the United States and Britain, as the food historian Andrew Smith recounts. In 1847, Harrison Crosby of New Jersey filled tin pails with whole tomatoes, boiled the pails, and sealed the top of each with a tin disk. Packed six pails to a box, Crosby sold them in New York and, to drum up business, sent samples to Queen Victoria and President James Polk, as well as to local newspapers, restaurants, and hotels. By the 1860s, one canner in New Jersey, employing thirty people, produced 50,000 cans in a single season. Consequently, by the 1870s, canning tomatoes had become big business in Pennsylvania and New Jersey, and by 1879 more than 19 million cans of tomatoes were being produced each year. Soon this figure had risen fourfold, and within a few years, more tomatoes were being canned than any other fruit or vegetable.

The first attempts at canning tomatoes in the newly unified Italy were nothing short of disastrous. A sample of tomato extract prepared according to the Appert method and shown at the first Italian national exhibition, held in Florence in 1861, was judged to have an "empyreumatic" taste. In other words, it had been burned. The hygienist Adolfo Targioni Tozzetti, the third of a Tuscan dynasty of naturalists and the author of a report dealing with diet and health, regarded the preservation of food as a national necessity, but he disapproved of using the Appert method of preservation.

At the same time, tomato ketchup had been proclaimed the national condiment of the United States. Early versions of tomato ketchup usually were simply boiled, reduced, and strained tomatoes to which a few spices had been added. It was used as a condiment with meat and fish or added to other sauces. Thus ketchup was not unlike the tomato coulis and purées found in the Italian cookbooks of the same period. Industrially produced ketchups, however, added ingredients like sugar, increasingly common in American cooking, as well as the more traditional vinegar and salt. After a shaky start, by 1899 the H. J. Heinz Company, based in Pittsburgh, Pennsylvania, had opened branches in London, Liverpool, Antwerp, Sydney, Bermuda, Mexico City, Toronto, and Montreal, on course to becoming the world's largest ketchup producer and constructing the world's largest preservation factory.

Ketchup, in the form of "Beefsteak ketchup," was then also made and sold by the Joseph A. Campbell Preserve Company, which manufactured canned tomatoes, vegetables, and, of course, soups. Its real success began in 1897 when the company developed a process for condensing its tomato and other soups. This, in turn, reduced production and shipping costs, and the lower prices were passed along to consumers. This, and a shrewd marketing campaign—which included recipe booklets and meal planners—made Campbell's soup a household name. And in 1912, Campbell's began to provide tomato farmers with seeds specially developed by the company.

Fresh tomatoes were being transported greater and greater distances by railway throughout the United States. When cultivation and consumption began to take off in the second half of the twentieth century, new varieties were developed to meet the increasing demand, with some developed specifically by and for the food industry. By the 1860s, American seedsmen had available more than twenty varieties, although we do not know exactly what they looked (or tasted) like. In Ohio between 1870 and 1893, Alexander Livingston developed or improved thirteen major varieties for the tomato trade. He named most of them after himself, such as 'Livingston's Marvel', 'Livingston's Magnus', 'Livingston's Paragon', and 'Livingston's Perfection'. Some of these varieties eventually found their way to Parma, Italy, suited as they were to the production of concentrate. In this guise, the tomato thus crossed the Atlantic a second time—before the finished product was exported back again.

State-run experimental farms, government agencies, and agricultural schools became interested in the fledgling tomato industry. Articles in the journal issued by the school at Portici, near Naples, suggest that beginning in the 1880s, agronomists became increasingly interested in the tomato, an interest that was as much technological and industrial as it was agricultural. In Brescia, the agronomist and priest Giovanni Bonsignori realized that for a farm to be a success, produce and processing had to be closely linked. In charge of the Istituto Artigianelli, an agricultural school for boys, Bonsignori began to grow tomatoes at the school's farm in Remedello Sopra. "Large-scale tomato cultivation [will start] the industrial processing of which is to be undertaken by a plant now being built in town, capable of processing any quantity of tomatoes." So promised Bonsignori in a speech delivered in January 1899. Later that year, Bonsignori wrote to the farm manager advising him on the use of fertilizers like nitrate of soda, ammonium sulfate, and bone phosphate on the 9 *piò* (7 acres) of land set aside for tomatoes. Nitrate of soda in particular, massive amounts of which were imported each year from Chile, had a long history in agriculture, and today, organic growers still are permitted to use it (figure 19).

Figure 19 "To fertilize your tomatoes, use nitrate of soda," recommends an advertisement published in the newspaper *Corriere della Sera*, April 19, 1913, 4. Note the large ribbed tomatoes favored by northern Italian growers, staked on sets of canes.

At first, things went well in Remedello Sopra. By 1900, the second year of its operation, the processing plant was producing some 200,000 cans of tomato concentrate for export, and other estate owners in the area were persuaded to grow tomatoes and send them to Remedello Sopra for processing. Then in 1904, tomatoes no longer appear as an entry in the school farm's accounts. Why was the experiment, which had seemed so promising, abandoned after just a few years? No reason is given, but as Bonsignori himself wrote in *La coltivazione del pomodoro* (1901), "Tomato cultivation, while possible in all the regions of Italy, is more successful in sunny places than in those areas infested by fog and cold." Perhaps Bonsignori had his native Brescia in mind.

One of the "sunny places" was Parma. The area was, and· is, justly famous for its ham and cheese. Less well known is its important place in the cultivation and preservation of tomatoes, which was largely the work of one man, Carlo Rognoni. Rognoni was a local landowner and professor of agronomy who decided to experiment with tomatoes after wheat prices fell in the 1880s. He succeeded where Bonsignori failed, overcoming local resistance to cultivating tomatoes and arguing that they could be successfully grown in the Po valley. So Rognoni inserted tomato cultivation into the preexisting cycle of crop rotation, along with careful irrigation and fertilization. Tomato cultivation in Parma also benefited from cheap seasonal labor and piecework, whose dramatic consequences we shall see in chapter 6.

On his own experimental farm, Rognoni tried out different tomato varieties—the preference was for larger fruits—and growing methods. This included the use of extensive staking to support the plants, which was able to dramatically increase yields. (Self-supporting, determinate plant varieties became available only after World War II, when they were imported from the United States.) The expansion of tomato cultivation throughout the province of Parma was accompanied by the development of tomato processing. Then, late in the century, the traditional drying of tomato paste into loaves—the *conserva nera*—was replaced by the bottling of tomato concentrate.

During the nineteenth century, bottling tomato concentrate remained a small-scale industry, both localized and marginalized. It was localized because Parma's producers were based at their production sites, close to the tomato fields, as many of the producers also were tomato farmers. The techniques used were essentially those of domestic production, albeit on a larger scale and using simple machines. The bottling was marginalized because at the International Exposition of 1889 in Paris, the few producers of *conserva* in the Italian pavilion found themselves dwarfed by the producers of other preserved foods. But tomatoes were not preserved at an industrial level until the twentieth century.

In 1891, however, Pellegrino Artusi offers us a tantalizing glimpse of the future. In his recipe for "saltless tomato preserves" (*conserva di pomodoro senza sale*), he refers to a new "procedure known as vacuum packing, in which you preserve fresh whole tomatoes in tin cans." He describes the production recently begun in Forlì (Emilia Romagna), which flourished at first. "Alas," Artusi concludes, "the treasury immediately slapped a tax on it, and the poor owner told me that he was thinking of closing down."

We have traced the development of the pasta-making industry and looked at tomato preserving. When did the two finally come together?

How and when the tomato-as-condiment was first put on pasta is a mystery. The first mention of using tomatoes in a pasta dish is actually French. In *L'Almanach des gourmands* (1807), Alexandre-Balthazar-Laurent Grimod de la Reynière recommended that in the autumn, tomatoes be substituted for the purées and cheese usually mixed into vermicelli before serving. He justified this practice by noting that "the juice of this fruit or vegetable (however one defines it) gives a rather agreeable acidity to the soups into which it is put, which is generally pleasing to those who have become accustomed to it." Using tomatoes in soups was a long-standing practice in Italy, as we saw in chapter 3, as Grimod would have known. But by Grimod's time, tomatoes also were consumed in Paris, as by then they had made the journey northward from Provence. Indeed, it is Provence that gives us the proverb "It is the tomato that makes a dish good" (*Poumo d'amour que'es bono viando*), advice that was about to be taken in various parts of Italy.

The marriage between pasta and the tomato is usually said to have taken place in Naples. *La cucina casereccia*, referred to earlier, has a recipe for *maccheroni alla napolitana*, in which the pasta is boiled in a meat broth in which tomatoes have been cooked. The cookbook is very enthusiastic about tomatoes, as we would expect, given its place of publication. Other uses of tomatoes, either fresh or preserved, are in meat stews and soups, with boiled meats and omelets, and as a filling for potatoes.

The recipe for *maccheroni alla napolitana* is not a tomato sauce. It was not until 1837 that Cavalcanti wrote that the secret of a successful dish of baked vermicelli with tomatoes (*timpano di vermicelli cotti crudi con li pomodoro*) was to make the tomato sauce dense, to cook the pasta just until firm, and to toss everything together in a pan. As for the accompanying tomato sauce, Cavalcanti wrote that whether it was made from fresh, dried, or preserved tomatoes, there was no point in describing its preparation, since everyone knew how to make it.

This recipe indicates that Cavalcanti was referring to an already established use, and not just in Naples. In fact, in the 1830s, the market for dried tomatoes was strong in Cagliari "because the sauce made from them [is] very good with *maccheroni*, which the middle and lower classes like so much." Likewise, Cagliari was then a renowned center for the manufacture of pasta. A recipe for "macaroni à la napolitana," combining pasta and tomatoes, first appeared in an American cookbook just a few years later, in 1847. By the 1880s, the tomato had been established as the condiment of choice for pasta for the peasants of the Campania region, and pasta itself had become a staple, as the findings of the Jacini Parliamentary Inquiry make clear. Even prisoners ate it, at least on religious holidays. On these occasions, prisoners in the southern Italian province of Capitanata were given *maccheroni al ragù*, the sauce made with tomato *conserva* and beef.

In Naples, the word for "tomato," *pommarola* (or *pummarola*), also referred, by extension, to a simple tomato sauce. In this guise, it even entered Tuscan regional cuisine, as *pomarola*. Although the Tuscans frequently used tomatoes, the sauce was evidently an import and is now regarded as a mainstay of "traditional" Tuscan cookery. A similar linguistic transfer took place in Sardinia. In Cagliari, tomato sauce became so common as a condiment while the island was under Piedmontese rule that it was simply referred to as *bagna*, the Piedmontese word for "condiment" or "sauce."

Not only was there a new use for the tomato, as a sauce for pasta, but by this time there also was a variety of tomato particularly suited to it. Dating from the first half of the nineteenth century, this small, egg-shaped variety was called the 'Fiaschetto' (also 'Fiaschella'), from the word for "small flask" or "powder horn." Italy's seed industry was in its infancy, in contrast to that in Britain, France, and the United States. But the 'Fiaschetto' tomato was one of three available as seed from the experimental farm in Naples in 1861. Twenty years later, several different varieties of related egg-shaped tomatoes were being grown in southern Italy, particularly around Naples. Most were made into conserves, according to *Les Plantes potagères*, by Vilmorin-Andrieux, France's premier nursery. The 1885, English edition of the book makes the first reference to a variety named in honor of Italy's King Umberto, when he made his first visit to Naples in 1878. The 'Re Umberto', or 'King Humbert', tomato was "distinguished by its rather peculiar form and appearance" (figure 20). Was the king pleased to be associated with this variety? The fruit had an oval shape but was flattened on the sides, "about the size of a small hen's egg," "of a very bright scarlet color"; grew "in clusters of from five to ten"; and was a most "extraordinary cropper."

The productivity of the 'Re Umberto' (and other egg-shaped varieties) made it attractive to growers around Naples. In addition, since it was a trailing variety, it was cheaper to cultivate, as it did not require staking. The city's high demand for tomatoes thus could be met. Naples also shipped out huge quantities of other fruits and vegetables. By the early 1890s, more than 30 tons of vegetables left Naples every day by

Figure 20 The 'Re Umberto' ('King Humbert') tomato (*left*) and its namesake (*right*).

rail, for the markets of Rome, Florence, Milan, and Turin. Most tomatoes, though, were still consumed nearby. The squat, bushy 'Re Umberto'
plants, with their numerous small, bright fruits, helped decorate country
areas, making them "appear to be . . . more gardens than fields," according to a report by the Ministry of Agriculture.

    One of the small ironies of history is that this most important vari-
ety of tomato was almost lost and is now a rarity eagerly sought by Italian collectors of "heirloom" seeds.

For King Umberto, it was a tomato, but for his wife, Queen Margherita,
it was a pizza. This was appropriate, since the tomatoes that first adorned
the *pizza Margherita* was the 'Re Umberto'. The story is that the queen was
tired of the French cuisine at court, so while she was visiting Naples in
1889, she brought a famous *pizzaiolo*, Raffaele Esposito, to the palace in
Naples, at Capodimonte. Esposito prepared three pizzas for her in the pal-

ace's pizza oven, which had been built for the Bourbon king Ferdinand II. One of the three pizzas was seasoned with tomato, mozzarella, and basil—the red, white, and green echoing the colors of the flag of the newly unified Italy.

Whether the pizza was invented for Queen Margherita or whether Esposito simply used an old recipe is unclear, but the latter is more likely. In any case, what is important is that a new culinary tradition was "invented" in the process. Many European culinary traditions, and the dishes they encompass, go back no further than this period, despite the patina of antiquity and timelessness subsequently attached to them, and the *pizza Margherita* is no exception.

As an "invented tradition," the *pizza Margherita* combined several elements: the populism of the new Savoyard monarchy at the expense of the vanquished Bourbons; the triumph of local, popular cooking over the imported French cuisine; and the Italianizing of a Neapolitan dish in the shadow of the Risorgimento, the movement to unify Italy. Today, in an age of greater regionalism, this tradition is harder to accept. Advocates of "southern pride" would like to see it renamed the *pizza Ferdinando*, in honor of the Neapolitan kings—father and son—who first developed a taste for pizza.

As is the case with pasta, the history of pizza began long before the arrival of the tomato. Pizza is a simple food, meant to be consumed on the move. In its soft form, it is reminiscent of the *nan* of India; in its hard form, it resembles the Berber bread of Algeria. In Italy, leavened or not, pizza had various names. By the nineteenth century, pizza had become an urban, plebeian food. The first references to pizza ovens, similar to bread ovens, where pizzas are publicly made and sold come from late-seventeenth-century Naples. 'Ntuono, where the Bourbon king Ferdinand II visited incognito, was already in operation by 1732. (It moved to the Port'Alba area of the city in 1738, and something like it is still there, although as a restaurant.) By the middle of the century, Naples had more than eighty *pizzerie*. Twenty of them had small tables outside for the convenience of their clientele. The *pizzerie* also supplied the peddlers who sold pizzas in the city's streets, carrying them on their heads in round metal containers to keep them warm.

The role of pizza in the life of the Neapolitan poor was described by Alexandre Dumas, who lived in Naples in 1835. In *Le Corricolo*, the author of *The Three Musketeers* and *The Count of Monte Cristo* (and much else) brought his Romantic novelist's eye to the city. Although Dumas may not be entirely reliable on points of fact, he did have a keen eye, and in matters of food, he considered himself an expert. Indeed, later in his life, Dumas wrote a six-volume dictionary of food, which is still in print.

Dumas says of pizza that although at first glance it appears to be a simple food, on closer examination it is actually quite complex. For six months of the year, pizza was the staple in the diet of the city's poor, as it was cheaper than pasta, which was reserved for Sundays. The price of pizza was determined by various factors, Dumas observed. First was its diameter. A two-*centesimi* size would feed a man, and a two-*soldi* size pizza would feed an entire family. The price also depended on the strength of the market, so it varied according to the freshness and price of the ingredients used, such as oil, lard, pork fat, cheese, tomatoes, or small fish. Pizza even could be purchased on credit and paid for later.

None of this impressed Carlo Collodi. In an anthology for Italian schools, *Il viaggio per l'Italia di Giannettino* (1882–1886), written just twenty-five years after Italy's unification, Collodi described pizza for his young readers:

> Do you want to know what pizza is? It is a flat bread of leavened dough, toasted in the oven, with a sauce of a little bit of everything on it. The black of the toasted bread, the off-white of the garlic and anchovies, the greeny yellow of the oil and the lightly fried greens, and the red bits of the tomatoes scattered here and there give the pizza an air of messy grime very much in keeping with that of the man selling it.

Beyond their use in pasta and pizza, tomatoes were fast becoming an indispensable element in the southern Italian diet. The nuns of Santi

Giuseppe e Anna in Monopoli (Bari Province) purchased tomatoes on sixty-three occasions in 1846—that is, nearly every day during the growing season—but the use of tomatoes was evident as well in other regions of Italy around this same time. Tomatoes and tomato purée took the place of pork fat as a condiment for staples like bread, polenta, and rice. Their increased consumption went hand in hand with their increased cultivation.

The place of tomatoes in rural Tuscany is well illustrated by Collodi's famous children's book, *The Adventures of Pinocchio*, first published in book form in 1883. One of its important themes is the juxtaposition of hunger and gluttony. Collodi creates a tension between money and eating, on the one hand, and poverty and hunger, on the other, which was not only the potentially tragic reality of Pinocchio's world but also the reality of Italian society at that time. In chapter 13, the Fox and the Cat are about to trick the wooden puppet Pinocchio into giving them five gold pieces, which Pinocchio was supposed to take back for his father, Geppetto:

> They walked, and walked, and walked, until at last, toward evening, they arrived dead tired at the inn of the Red Crawfish. . . . Having gone into the inn they all three sat down to table: but none of them had any appetite. The Cat, who was suffering from indigestion and feeling seriously indisposed, could only eat thirty-five mullet with tomato sauce, and four portions of tripe with Parmesan cheese; and because she thought the tripe was not seasoned enough, she asked three times for the butter and grated cheese. . . . The one who ate least was Pinocchio. He asked for some walnuts and a hunch of bread and left everything on his plate.

Some time later, Pinocchio jumps into the sea, only to find himself in a fisherman's net. Pinocchio explains to the fisherman that he is not a fish to be eaten, but a puppet. The fisherman replies that he has never caught a "puppet fish" and asks how he would prefer to be cooked: "Would you like to be fried in the frying pan, or would you prefer to be stewed with tomato sauce?"

In Tuscany, the tenant farmers (*mezzadri*) used tomatoes sparingly in soups or with stale bread in their *pappa col pomodoro*. Essentially a bread-and-tomato soup, *pappa col pomodoro* became known throughout Italy thanks to a children's story first serialized in 1907/1908, Vamba's *Il giornalino di Gian Burrasca* (which has never appeared in English). "Long live *pappa al pomodoro!*" becomes "Johnny Hurricane"'s motto as he rebels against the dull diet of rice soup at the boarding school where his parents have sent him in an attempt to teach him how to behave.

Bread was still a basic foodstuff in Italy, an essential element of the daily diet. More often than not it was consumed stale, since in rural areas bread was baked in communal ovens only every week or two. Another humble dish of the Tuscan region was *panzanella*, in which pieces of stale bread were moistened in water or vinegar and broken into crumbs. To this was added whatever seasonal vegetables the *mezzadro* could grow in his own kitchen garden: onions, carrots, cauliflower, greens, cucumbers, and, as an innovation, tomatoes. The innkeeper Bicchierai called *panzanella* "a poor man's dish, but tasty and pleasing to gentlemen, priests and government officials, too." It is a useful reminder to know that something created out of necessity and poverty might also be considered "good" at the time. Like the *acqua-sale* popular in parts of southern Italy, which it resembles, *panzanella* today is touted as a crucial part of the "Mediterranean diet." Both are now inconceivable without tomatoes.

The peasants of Lazio added tomatoes to their thick vegetable soup, minestrone, or used it to make a "mock sauce" (*sugo finto*) for beans or lentils. Tomatoes even appeared farther north. In *Le barufe in famegia* (1872), the Venetian playwright Giacinto Gallina has his lower-middle-class family arguing over whether to have rice with tomatoes for supper. In Udine (Friuli), a recipe for tomato sauce was included in a handwritten collection of 182 recipes compiled in the early nineteenth century by the noble Caiselli family. A similar collection compiled by Lucia Prinetti Adamoli in Varese (Lombardy) and dating from the mid-nineteenth century contains a recipe for "tomatoes à la provençal." She copied the recipe from Giovanni Vialardi's *Trattato di cucina* (1854), but it originated decades earlier in Paris and also can be found as Vincenzo Corrado's *pomidori alla Napolitana*. The recipe is for baked whole tomatoes stuffed with grated cheese, bread crumbs, and parsley and moistened with broth or sauce.* Italy's Jews were eating stuffed tomatoes too, with rice and bread crumbs, according to an anonymous poem written for the celebration of Purim.

It is in the first explicitly "Italian" cookbook that we find the best examples of the spread of tomatoes, especially in the form of *conserva*. This was Pellegrino Artusi's *La scienza in cucina e l'arte di mangiar bene*, first published in 1891. It was Artusi's mission to create an Italian cuisine, thirty years after Italy's unification. The recipes in the book are from Tuscany, Emilia Romagna, and other regions, as well as a mixture of high and low cuisines. Artusi gives two recipes for tomato *conserva*, one savory and one sweet. He stresses the great usefulness of the *conserva*, commenting that "if the tomato were rarer, it would cost at least as much as truffles." For example, in Artusi's recipe for tripe, a poor man's dish, the tripe was to be cooked in butter to which a meat sauce would be added or, "if you do not have that, tomato sauce."

In his cookbook, Artusi distinguished between *sugo di pomodoro*— tomatoes cooked and strained, with the possible addition of celery, parsley, or basil—and *salsa di pomodoro*. He told this story about the *salsa*:

There was once a priest from Romagna who stuck his nose into everything, and busy-bodied his way into families, trying to interfere in every domestic matter. Still, he was an honest fellow, and since more good than ill came of his zeal, people let him carry on in his usual style. But popular wit dubbed him Don Pomodoro

---

* The Adamolis liked the recipe enough to grow tomatoes on their estate in 1864. Because their son Giulio had taken part in Garibaldi's invasion of Sicily, as one of "the Thousand," it is possible that he learned to like tomatoes there and brought some seeds home with him, thus helping unite the new country in a culinary way, too.

(Father Tomato), since tomatoes are also ubiquitous. And therefore it is very helpful to know how to make a good tomato sauce.

This is followed by brief recipe for tomato sauce.

Pellegrino Artusi's Tomato Sauce (*Salsa di pomodoro*)

Prepare a *battuto* with a quarter of an onion, a clove of garlic, a finger-length stalk of celery, a few basil leaves and a sufficient amount of parsley. Season with a little olive oil, salt and pepper. Mash 7 or 8 tomatoes and put everything on the fire, stirring occasionally. Once you see the sauce thickening to the consistency of a runny cream, pass it through a sieve and it is ready to use.

This sauce lends itself to innumerable uses. . . . It is good with boiled meat, and excellent when served with cheese and butter on pasta, as well as when used to make the risotto described in recipe 77 [*risotto coi pomodori*].

    Pellegrino Artusi, *Science in the Kitchen and the Art of Eating Well*, trans. Murtha Baca and Stephen Sartarelli (Toronto: University of Toronto Press, 2003), 121.

Artusi also provides two recipes for "Neapolitan-style macaroni" (*maccheroni alla napolitana*), and the second is now a mainstay of Italian cuisine: classic comfort food. Artusi's first recipe is for a gently simmered, meat-based sauce for pasta using tomato *passata*, with the meat from the sauce served on the side. The meat should be a mixture of beef and pork, a "hodgepodge of spices and flavors," which Artusi finds pleasing. He claims to have obtained this recipe from a family in Santa Maria Capua Vetere, just north of Naples. Along with millions of Italians, Neapolitan-style macaroni, or something much like it, traveled across the Atlantic, to become a classic of Italian American cooking.

5

"Authentic Italian Gravy"

By 1940, the first mass migration of millions of Italians was over, but it had led to what is still one of Italy's most prosperous industries, the growing of tomatoes and the preparation of tomato products. More than that, Italian food has spread and become popular the world over. What Remigio Baldoni wrote in *Il pomodoro industriale e da tavola* (1940) is true today:

> Tomato paste is the most typical condiment of Italian cuisine; it gives color and a characteristic flavor to many of our traditional dishes. The Italian forced to live far away from his homeland, wherever in the world he sets his table, rejects every kind of cooking in order to establish his own, the simple but tasty cooking of his native land. And more than anything else he does not give up his traditional dish of macaroni with tomato sauce. The influence of our fellow countrymen living overseas has thus given rise to a thriving flow of exports from Italy, which, beginning with the period of great emigration onward, has reached imposing levels. Cheese, pastas, and tomato preserves are the classic products of our industry that have traversed the oceans and continents, finding a place not only on the tables of Italians but on those of many foreigners as well.

Argentina, Brazil, and Chile are South America's largest producers of tomatoes, but the tomatoes did not come from Mexico, where they originated. Rather, they came with the Europeans, notably the Italians, in the late nineteenth century, so they had to cross the Atlantic Ocean twice. Italian immigrants in South America brought with them a taste for the tomato paste used in cooking. At first, beginning in the early twentieth century, large quantities were imported from Italy, but by World War II, a local industry had developed.

The story of the tomato in the United States, the other main destination for millions of Italians, is slightly different. Here the tomato had already found success as a condiment, in the form of ketchup. Tomatoes were extensively grown in states like New Jersey and Pennsylvania, and American seedsmen were at the forefront of developing many new tomato varieties. Companies like Heinz and Campbell's were becoming international concerns, but Italians still managed to have a profound influence on the cultivation and consumption of tomatoes in the United States. At first, tomato preserves were imported from Italy to meet the demands of the Italian "colonies" spread throughout the United States. Then, as in South America, local production began to compete with the imports, with help from the government.

The massive displacement of people that historians have called the "Italian diaspora" altered the lives of millions. Between 1876 and 1945, more than 9 million Italians crossed the Atlantic Ocean to the Americas, most of them arriving between the 1890s and the 1920s. As many as half that number then returned to Italy or migrated repeatedly. These migrants' effect on the American diet and culture, particularly in regard to the tomato, is the subject of this chapter.

Typical of these immigrants is a character in a novel by the Italian journalist Bernardino Ciambelli. The novel is *I misteri di Mulberry* (The Mysteries of Mulberry Street, 1893), and the character is Vittoria Ruiz, a Neapolitan, who settled in New York City:

Vittoria had come to America, without even thinking of the dangers she might meet. . . . When Vittoria stepped into Mulberry Street she looked around her with wonder, as if waking from a long, deep sleep. That street reminded her of one in the beautiful city which had seen her birth; she seemed to be in one of the popular quarters of her delightful Naples. . . . Her companion put an end to those imaginings, brusquely telling her that they had to hurry home, where her man must have been in a fury for having missed his mid-day meal. . . . "Here we are, my dear," the old woman said all of a sudden, "it may not be a palace, but it will have to do." It wasn't a palace, quite the contrary, for the low house of two short stories, its dark façade the color of clotted blood, had the appearance of a hovel. At the ground floor were two shops with dirty windows, broken in places . . . and under those shops a passage, leading down to a basement with a crude sign, on which was scrawled "Ristorante Italiano, Maccheroni alla pumarola."

In melodramatic fashion but with an attentive eye for detail and color, Ciambelli conveys the atmosphere of New York City's Italian "colony." For the Neapolitan immigrant, everything is at once the same and yet completely different. It is a squalid version of home, complete with a self-styled Italian restaurant serving that quintessentially Neapolitan dish, pasta with tomato sauce. As a journalist and a native of Lucca, Ciambelli was twice a foreigner, to both the United States and the largely southern, working-class Italian community of which he became a resident.

Other Italian writers in the United States turned to pasta when they wanted to add a note of rustic southern Italian color to their New World settings. For example, in his short story "Peppino" (1886), set in New York City, the Veronese teacher Luigi Donato Ventura has his young shoe-shine boy Peppino devour "the traditional dish of macaroni," as if to underscore the boy's Neapolitan origins. This reference to pasta may not seem so remarkable to us, so closely identified is pasta with Italian

American cooking. But at that time, such a reference was remarkable, for a couple of reasons. First is because pasta with tomato sauce, as a dish, was still a novelty here when the Italian mass migration began, and second is because most of the immigrants living in New York were not from the city of Naples. Although by now, tomatoes were a standard ingredient in their diet, pasta was not, at least not yet. But it became so in America. Paradoxically, one of the means by which pasta became identified with Italian American culture, for both Italians and Americans, was through Italian restaurants.

The change in diet began on board ship. Peasant migrants traveling across the Atlantic may have eaten better than at any time in their lives. If an Italian ministerial decree of 1904 stipulating the on-board diet of emigrants—and it's a big "if"—is to be believed, then they ate quite well. Each week, they were to be given 8.8 pounds of wheat bread, 3.3 pounds of beef, 2.2 pounds of pasta, and 7.4 pints of wine. They also were to receive rice, salt cod, tuna, anchovies, cheese, potatoes, beans, lentils, peas, onions, olive oil, tomato *conserva*, coffee, and sugar.

For the migrants, it must have seemed to good to be true (and it probably was). Leonard Covello, who became a high-school teacher and administrator, remembered life in Avigliano (Basilicata), which he had left at the age of nine, as one characterized by deprivation. "There were times when there was no food in the house," he recollected. "Then we bolted the door and rattled kitchen utensils and dishes to give the impression to our close neighbors that the noonday meal was going on as usual."

The contrast was reinforced on the immigrants' arrival. Life may have been daunting and the toil unremitting, but the food was likely to have been more abundant than what they were accustomed to. The most radical change was the easy access to meat. In Brazil and Argentina, Italian immigrants were surprised to find that meat was more common and cheaper than vegetables, a complete reversal of traditional food hierarchies. It also was the realization of one of the peasants' strongest aspirations. "In America I ate meat everyday." These were the words of a peasant from Nicastro (Calabria) who had worked in Pittsburgh, identifying what was for him the most novel aspect of his American experience in the early twentieth century. Even the poorest could afford to eat meat, like the New York shoe-shine boy Rocco Corresca. His family had "very little money, and our clothes were some of those that we found on the street. Still we had enough to eat, and we had meat quite often, which we never had in Italy." A peasant from Cuneo recounted the main benefit in an otherwise difficult situation: "The only thing that wasn't lacking was meat, so [the immigrants] didn't suffer from hunger. In Argentina meat was like polenta is for us." In other words, meat was a staple. Workers in the rural areas of the Pampas even became tired of a diet of "pasta, rice, meat, and yet more meat," in the words of one Piedmontese.

Nevertheless, as the historian Simone Cinotto has pointed out, the first phase of immigration was characterized by frugality. Saving money was paramount, whether it was for the return journey or in order to become established. When a New York medical doctor examined two Sicilian women in 1906 and found them to be seriously anemic, he asked them about their diet. The two women confirmed that "you come to America to work, not to eat." Old habits persisted. Italian peasants, with no state aid, lacking qualifications, and shouldering the debts taken on in order to emigrate, kept close to the only world they knew. This was partly for reasons of cultural identity and partly because of the focus on saving. They achieved this, in the words of the *Bollettino dell'emigrazione* (1911), "by eating little, living primarily on pasta, cheese, [and] vegetables, usually of inferior quality if not altogether spoiled" (figure 21). They often thought of themselves as temporary migrants, saving money only to finance their return home. Or they saved to send money home or to pay for passage for family members. Only after the immigrants became more firmly rooted, with more secure jobs, and settled down with families of their own, did their diet improve.

The dietary frugality of the early Italian immigrants in the Americas was thus conditioned by circumstances, but what might have been perceived

Figure 2 I An Italian family eats supper on the Lower East Side, New York City, 1915. Gelatin silver print by Lewis Wickes Hine, from a series of photographs of social conditions. (Photograph Collection, Miriam and Ira D. Wallach Division of Art, Prints and Photographs, New York Public Library, Astor, Lenox, and Tilden Foundations)

as a sign of their frugality by the host community might also have been a matter of choice. Their preference, and desire, was for vegetables and fruit, which already were a prominent part of the immigrants' diet, especially if they came from southern Italy. Financial considerations and taste preferences therefore came together in the domestic vegetable gardens that Italian immigrants planted as soon as they could. Since most were of peasant stock, it was natural for them to keep a vegetable garden, raise an animal or two, gather wild plants, and preserve food.

Later, when social workers and women's organizations recommended planting "victory gardens" after the United States entered World War I, they found that Italians were way ahead in home-cultivated and -preserved food. Backyard gardens also represented a bittersweet link with the world the immigrants had left behind. Joe Vergara recalled his child-hood house in Queens, New York:

> The back yard was all Pop's—his little patch of Italy on Long Island. He spent every spare minute working in his tiny plot. . . . Rows of tomato and pepper plants in the center, tied neatly to individual stakes; beans along the fence, trained on strings, some-times a few eggplants. . . . In the north corner of the yard, Pop built a grape arbor and under it he placed two benches. This was his retreat. Sitting under his maturing grapes—wine-to-be—he could forget the growing pile of bills in his shop, the overdue payment on the furniture, and his fading prospects of returning to Calabria with a trunkful of gold.

For the many Italian immigrants working in factory jobs, their household vegetable plots also served to reconnect them to the land. Some immigrants even were able to put their Italian experience of grow-ing and marketing produce to good use in America. As early as 1880, 12 percent of New York's Italians were peddling fruits and vegetables, which was the most important autonomous occupation for immigrants. The Italian peddler was a common enough figure in North American towns and cities that a reading lesson in a Canadian school primer showed a picture of a smiling mustachioed man named "Mr. Conti" selling fruit. Although it was backbreaking work, it was a start. "In New York," wrote R. F. Foerster in 1919, "many [Italian] men have begun with a pushcart, then got the privilege of a market stand, then a concession to sell garden produce in connection with a grocery store, and finally have set up a shop on their own." By the 1920s, more than half the fruit and vegetable stores in Montreal and Toronto were owned by Italians, and most smaller cities had Italian fruit and vegetable markets.

The range of produce on display at Italian greengroceries was a source of wonder to native New Yorkers. "The variety of salad greens is remarkable," wrote Sophonisba Breckinridge in *New Homes for Old*

(1921), a study of the acculturation of immigrants in the United States: "More Swiss chard, mustard, dandelion leaves, endive, squash blossoms and leaves, escarole, are to be seen in one little Italian store than in half a dozen American markets." Italian greengrocers in New York introduced the city to a range of vegetables now taken for granted, as the names of some attest: zucchini, asparagus, eggplant, celery, broccoli, fennel, green peppers, escarole, and rapini.

Italian food retailers were part of a developing food chain that stretched from market gardeners to consumers and included wholesalers and restaurateurs. Italians played a crucial part in every link of the chain. Networks of immigrants formed as result of the establishment of labor gangs and boarding arrangements. The food that their "bosses" provided was the logical extension of the frugal meals offered to day laborers and farm hands (*braccianti*) on the agricultural estates of southern Italy. For breakfast, immigrant labor gangs had coffee and stale bread; for lunch, bread and mortadella, or sausage and onion; for supper, pasta or rice (with tomato sauce), potatoes, or beans. Such a diet favored the sale of Italian food imports. At the beginning of the twentieth century, immigrant workers laying tracks for the Canadian Pacific Railway depended on their "bosses" in Montreal to supply Italian foodstuffs to their labor camps. Moreover, at their work sites, Italian laborers went out of their way to buy from fellow Italians, even when this might cost up to five times the city price. In Argentina, this preference was made easier by the fact that the main food warehouses (*almacenes*) were run by Italians.

As the Italians settled, the chain lengthened. At one end were Italian market gardeners selling produce to their countrymen. Quite often, they delivered door-to-door, selling bushels of tomatoes directly to eager customers. Progressing from market gardening, Italians became wholesale distributors of fruits and vegetables. In New York City, the business was lucrative enough to attract the attention of the local Mafia. One gangster, Ciro Terranova, "the Artichoke king," took over the sale of artichokes in the city's open-air markets, forcing stall holders to buy from a designated supplier. The racket was not broken up until the mid-1930s.

The bushels of tomatoes delivered to awaiting householders were put to good use. Making homemade tomato paste (*conserva*) was, for many immigrant families, partly a symbolic link to the town left behind, partly a matter of taste preference, and partly good economic sense. The two methods most often used dated back to at least the 1860s in Italy. In New York City in the early twentieth century, strained, peeled, and seeded tomatoes were put to dry in the sun on tenement rooftops and stirred frequently so that they evaporated and became concentrated. In the Greenbush area of Madison, Wisconsin, drying boards were common (figure 22). "Fresh tomatoes were strained and cooked until the water disappeared," one local woman recollected. "They were set on wide tables and sun-dried until thick and only small amounts remained." It was "tomato board time" for another woman, the event memorable enough to earn its own label. "Tomato board time wasn't our favorite. . . . My four sisters and I were expected to stand in the heat, shoo the flies and support the boards on our shoulders that were resting at an angle off the backyard fence."

The second way to make tomato paste produced something similar to a bottled *passata*. The following account comes from Nancy Verde Barr's survey of Italian American cooking in the Little Italy of Providence, Rhode Island. Likewise, her American-born informant viewed it as a chore, but she nevertheless offered a detailed account of the process, including the soda bottles used as containers:

My mother put up her own tomatoes and that was a chore we had to help with. She had a big galvanized tub down in the basement and she used to put a nice clean cloth over it. Then she had a strainer which was like a flat tray with raised sides and holes in it. We would squeeze the tomatoes on that so all the skins and seeds would remain on top and all the good stuff would come out on the bottom. Then she would pour it into this big cloth and all the water would drain out; then we would scoop that and put that in bottles. We used to squeeze baskets and baskets of tomatoes in this way. We weren't allowed to touch anything while doing this—if you

Figure 22 Making tomato paste the Sicilian way, in the Greenbush area of Madison, Wisconsin, mid-1920s. (From Catherine Tripalin Murray, *A Taste of Memories from the Old "Bush"* [Madison, Wis.: Italian-American Women's Mutual Society, 1988], and Anthony Fiore)

had an itch you had to use your elbow. It was a tradition for girls, if they were having their periods at the time, they weren't supposed to do this. Sometimes if you didn't feel like squeezing tomatoes, you would pray that you got your period.

Making and bottling tomato paste, whatever the method used, required space, and in many North American houses, this gave rise to the "second kitchen." This work space was located in the basement, a shed, or the garage, thereby freeing up the upstairs kitchen to assume a more formal role. The second kitchen showed that the family had "arrived." Cooking for a large number of people and preparing food for storage could even be said to have had an architectural impact, altering the way that spaces were used.

Many Italian immigrants preferred buying imported Italian tomato paste in cans. In any case, the tomato paste was stored for use throughout the year. Being able to put food aside for future needs was a peasant's eternal aspiration. In the New World, it was a symbol of plenty. In *The Fortunate Pilgrim* (1964), Mario Puzo's evocation of life in the tenements of New York in the 1920s, we learn that the Angeluzzi-Corbi family used the back of their hallway for extra storage space: "The icebox, a bureau, countless cans of tomato paste, and boxes of macaroni were stacked against the wall."

Many Italians in the Americas bought imported tomato paste. In the first three decades of the twentieth century, when the Italian production of tomato preserves reached a peak, Italy exported much of what it produced. Although internal demand was minimal, as Italy still was largely a peasant society, the resulting surplus led to an abundance of cheap exports, and foreign demand spiraled. The main importers were mainly countries with large populations of Italian immigrants: the United States, Argentina, Brazil, and France.

At the start of the twentieth century, eleven factories were producing tomato preserves in the Naples area alone, employing 670 people. The city's authorities had assigned a special warehouse for storing the canned tomatoes awaiting export. Tomato paste was produced throughout Italy, from Milan in the north to Siracusa in Sicily. Italian exports of tomato preserves, mostly in the form of concentrate or paste, increased steadily in the twenty-five years between 1905 and 1929, from 11,473 tons in 1905 to a peak of 150,698 tons in 1929. An agronomist in Parma called it "tomato fever," as local farmers and producers shifted to tomatoes in order to cash in on the growing international demand. By 1910, there were thirty-six plants in the Parma area, providing work for three thousand people during the fifty days of the canning season. Associated industries—the manufacture of cans, labels, and crates—also benefited. Only World War I brought a momentary halt during this period.

The greatest importer of Italian tomato paste was also the country with the greatest number of Italian immigrants: the United States. From 1905 to 1929, the United States consumed close to half the Italian exports (46.5%). Argentina, with fewer southern and more northern Italian immigrants, imported less than one-tenth (8.4%) of Italy's tomato paste. Argentina also began to produce some of its own tomato paste in 1910, when the government awarded a certificate of merit to an engineer and industrialist from Parma, Tito Manzi, for helping set up a canning factory there. The Italians in America were the main consumers of manufactured tomato paste, buying more and more imports as their purchasing power grew. Sometimes, however, whole consignments of canned Italian tomato paste were rejected for admission into the United States because government scientists objected to additives in the paste. Imports were also interrupted during World War I.

Much as they are now, food additives became a contentious issue. National governments, producers, and regulators all disagreed on what should be considered permissible. Consumers had little to say. As early as 1863, the Neapolitan doctors Achille Spatuzzi and Luigi Somma worried about the "heterogeneous substances" and colorings added to the tomato paste sold to the poor. Food adulteration was a problem before the government regulation of additives, and, indeed, it still is a problem. The adulteration of preserved food products, however, was sometimes tolerated as long as the ingredients were not considered harmful.

According to Renato Rovetta's *Il pomodoro* (1914), a guide to tomato cultivation and production, a whole range of substances could legally be added to canned tomato paste and other preserved foods. The permitted colorings included the aptly named "tomato red," which was made from coal tar and described as "resistant" and soluble in water, 0.05 cup of coloring to about 26.5 gallons of tomato paste. Several preservatives were permitted as well, from various kinds of acids (the most common being the disinfectant salicylic acid) to formaldehyde and more common additives like table salt, vinegar, and sugar. Beginning in 1890, Italian legislation took a "liberal" approach, allowing additives in foodstuffs as long as they were harmless and served only to help sell the product and facilitate its use. The only exception was that the additives could not fraudulently interfere with the volume or weight of the product, nor could they be used to camouflage poor quality.

In the United States, manufactured tomato ketchup, then a rising star in the food industry firmament, was likewise widely adulterated. Coal tar was used to produce a bright red coloring, and starch was added as a thickener. Abuses were not prosecuted until the passage of the Pure Food and Drug Act in 1906. Because the American legislation was much more extensive and much more focused than the Italian, tomato paste had to contain what was listed on the can, nothing else.* Therefore, Italian producers who wanted to export to the United States had to adapt or risk having their shipments chemically analyzed and perhaps rejected.

Rovetta was clearly aware of the two different—not to say irreconcilable—approaches adopted by Italy and the United States. As the son of the industrialist who had created and made the first machines for the industrial manufacture of tomato paste, he was all but silent on the matter of adulteration, although he admitted that "in the past" there had been "cases of the importation [into the United States] of not very good or defective tomato paste." But he certainly did not suggest that manufacturers abandon their use of additives, which were, after all, legal and made the product more attractive to consumers. As Rovetta pointed out, 15 to 20 percent of Sicilian tomato paste was added salt, why complain? It only meant that cooks had to use less of this highly taxed commodity in their cooking. No doubt there were some "unscrupulous" producers, but

* The only additive permitted today are calcium salts, which help the tomatoes remain firm.

the solution was to buy a reliable and well-known brand. It is no wonder that so many Italians preferred to make their own tomato paste!

Even when the producers of tomato paste did break the law, it was up to the local authorities to prosecute them, and in areas like the Naples hinterland, which relied on the food industry, this was not easy.

We should not be surprised that the United States imported so much tomato paste, despite the restrictions. After all, the demand was there, and the country did not yet manufacture much of its own. Instead, the great surprise is the identity of the second main importer of Italian tomato preserves: the United Kingdom, which consumed more than one-fifth (22.5%) of all exports. The volume of British imports started low but increased exponentially during the early twentieth century, in line with Italian production. Furthermore, instead of declining during World War I, British imports actually peaked between 1915 and 1917. Indeed, in 1917 Britain almost single-handedly saved Italy's tomato industry, buying nearly three-quarters (73.3%) of its exports.

The Italian product had a virtual monopoly in Britain, mirrored by the Italian reliance on British tin. This was the reverse of the situation in the United States and Argentina, whose own tomato production was on the rise, eventually leading to a decline in imports. Partly as a result of this, British demand was able to affect production strategies in Italy. In fact, and this may come as a shock, the British were indirectly responsible for the introduction of the 'San Marzano', one of Italy's great tomato varieties and the pride of southern Italian growers.

In the early twentieth century, the British were quite happy to import preserved tomatoes in the form of paste or concentrate, most of it from the Parma region, to be used in soups and sauces. But the British had really wanted was fresh tomatoes, which was one of the first "vegetables" to be appreciated raw. The British also liked fresh tomatoes cooked, especially baked or grilled. Accordingly, the cultivation of tomatoes had spread from the large country estates to become one of the items most often found in home gardens. The cultivation of tomatoes spread with the growth of the suburbs and the increasing availability of greenhouses to protect British tomatoes from the inclement weather. At the same time, the Channel Islands were supplying Britain's urban markets. But so far, there was no way for whole tomatoes to be enjoyed beyond the short growing season. An attempt was made to market canned crushed tomatoes in Britain, an idea of the fashionable French chef and author Georges Auguste Escoffier. Manufactured by the Caressa firm in Escoffier's native city of Nice, they did not catch on, however.

In southern Italy, it turned out that the smaller, egg-shaped tomato varieties that were the most popular here also were suited to being canned whole. The production method was simpler than that required for tomato concentrate. The tomatoes were washed and scalded by machine and then peeled before being sterilized and canned. The fruits had to have a good solid pulp, with little core, and, after being peeled, would not lose their shape in the can. The larger, ridged varieties, favored in central and northern Italy for tomato concentrate, did not have these characteristics. Ferruccio Zago's *La coltivazione industriale del pomodoro* (1913) notes the use of the 'Re Umberto' for tomato paste but added that it was now grown primarily "for the so-called *pelati*, which are tomatoes prepared in their natural state with the skins removed."

This development led to a major shift in production and consumption. During the 1920s, the exports of Italian tomato concentrate gave way to the cheaper canned whole tomatoes, making Parma's loss into Salerno's gain. At the beginning of the decade, two-thirds of the preserved tomatoes were exported in the form of paste, but by the end of the decade two-thirds were exported as *pelati*. The production of canned whole tomatoes was initiated in order "to respond better to the taste of the Anglo-Saxon countries, which prefer this product to overly concentrated sauces," according to Baldoni's *Il pomodoro industriale e da tavola*, also noting that "this production, truly delicious, conquered the tastes of connoisseurs within the space of a few years." The result was the spread of the cultivation and production of tomatoes. British and American imports more than doubled during the 1920s over the previous decade.

Producers for the British market labeled their cans of "Naples peeled plum tomatoes" with brand names like the English-sounding "Goodwood Cup."

The tomato variety that made all this possible was what Baldoni's manual called "a recent cross" between the 'Re Umberto' and 'Fiaschetto' varieties. This was the 'San Marzano', named after the town near Salerno where it was first cultivated. In just a few years, the 'San Marzano' had become the main variety used for canned whole tomatoes, and by the mid-1920s Italy's two largest seed companies, Ingegnoli (Milan) and Sgaravatti (Padua), were selling 'San Marzano' seeds.

The production of tomato concentrate did not stop, of course, during the first thirty years of the twentieth century. Besides experimenting with new technologies and new tomato varieties, industrial producers were interested in finding two particular qualities: pulp and juice. Consequently, new varieties like the 'Ingegnoli' were being grown alongside the traditional large, segmented varieties like the 'Large Red' and the 'Genovese'. Zago's manual on industrial tomato cultivation describes the 'Ingegnoli' as a cross between two American varieties, the 'Ponderosa' and the 'St. Louis', and recommends it as particularly well suited for *conserva*. This is a reminder that the cultivation and production of tomatoes in Italy did not exist in a vacuum. Traditional varieties were being crossed with newer American imports in an ongoing search to maximize taste, productivity, and disease resistance.

The trend toward ever larger fruits, which was such a preoccupation of American seedsmen and producers (like Campbell's and Heinz) until the 1920s, was not a major concern in Italy. The boast that a single tomato was enough to fill a can seems to have fallen on deaf ears. For instance, the "large fruits" and high productivity of the 'President Garfield' variety were noted in Zago's manual, but this tomato was also rated as "not one of the best." Nonetheless, the varieties used in Italy and the United States showed signs of converging, at least in terms of the characteristics considered important. In 1926, Heinz abandoned the large beefsteak tomato for a smaller, fig-shaped variety for its ketchup. This tomato was high in pulp, viscosity, and pectin. Even more important, all the fruits ripened at the same time, which suited mechanical harvesting. It was a variety specifically adapted to both the farm and the factory. The resulting new, denser paste was perfectly suited to ketchup for emerging American foods like hot dogs, french fries, and hamburgers.

Curiously, several of the varieties most appreciated by Italians today for consumption raw in salads were first developed for the canning industry and for export abroad. Indeed, Italy still exports around three-quarters of its processed tomatoes.

If some tomato varieties were crossing the Atlantic from the United States, others were making the opposite journey, literally following the Italian immigrants. The 'San Marzano' was one of these.

The period from 1900 to the start of World War I was the golden age for Italian food imports in the United States. Italian immigrants preferred imported tomatoes, pasta, oil, cheese, and preserved meats over locally made goods, even when these were available and even when they had to pay a 40 percent tariff on Italian tomato paste. Canned whole tomatoes, not yet a common export from Italy, had a smaller tariff of 15 percent. (Finally, in 1978, Italian tomatoes had a tariff of their own.) This was not the first time a tariff had been imposed on foreign tomatoes. In 1883, the United States Congress levied a duty of 10 percent on imported "vegetables" at a time when fresh tomatoes were being imported early in the season from British colonies in the Caribbean. A New York importer, John Nix, objected that since tomatoes were a fruit, they should be exempt from this tariff. Nix even took his case to the Supreme Court. But he lost when the Court ruled that botanically speaking, tomatoes might be a fruit, like cucumbers and squashes, "but in the common language of the people, whether sellers or consumers of provisions, all these are vegetables." Tomatoes were served at dinner as part of the main course, and not the dessert, the ruling concluded.

When World War I brought a virtual halt to imports of processed tomatoes from Italy, local producers stepped in to fill the gap. Some of

New Jersey's and California's tomatoes were now destined for tomato paste, which was manufactured in the "Naples" style, sometimes labeled *salsa* in Italian. This tomato paste was less concentrated than the tomato paste made in Parma. In addition, *salsa* might contain other ingredients such as onions, carrots, and herbs, so that it could be used directly on pasta, according to one Italian producer. But American manufacturers did not limit themselves to tomato paste. To meet the ever increasing demand for Italian-style canned whole tomatoes—which were smaller, darker, sweeter, and less juicy than their American counterparts—local manufacturers added tomato paste to their canned California tomatoes. The product was developed mainly to meet the tastes of Italian Americans, and during this period, California wholesalers sold as many as 1 million cases of these canned tomatoes to Italian residents in and around New York City. When imports from Italy resumed after World War I, however, these local producers ran into serious difficulties. Although their product was slightly cheaper than the imported one, it was perceived as inferior.

The production of tomato paste and canned tomatoes in the United States became successful only in the 1930s when several factors came together. First was the Smoot-Hawley Tariff Act of 1930, which raised the tariff on imported tomato products to 50 percent. This was followed a few years later by a temporary boycott of Italian goods after Italy's invasion of Ethiopia in 1935. Production in Italy then fell, owing to Fascist policies aimed at self-sufficiency in grain and the slowdown in international markets. At the same time, the American products improved in both quality and quantity, the result of efforts by producers. To do this, they brought over qualified personnel from Italy and moved most of the cultivation and production to California, thereby expanding that state's already flourishing canning industry. Finally, Italian importers themselves turned into producers, putting their knowledge of the American market to good effect.

Two examples illustrate these developments. California's River Bank Canning Company was founded in 1923 by Lorenzo Zerillo, who, according to the company lore, arrived in the United States with a few varieties of tomato seeds in his pocket. The company then become one of the first producers of tomato paste, taking full advantage of the production and marketing strategies employed by the rapidly expanding American food industry. River Bank bought its tomatoes from farmers under contract to the company and kept its labor costs low by employing seasonal Mexican workers, mostly women, for the canning. It sent most of its output by train to its New York office, where it was sold directly to the major wholesalers and large supermarket chains. In New York, the tomato paste was sold under the very Italian-sounding label of "Madonna Tomato Products." The company managed to combine a traditional, "ethnic" image with up-to-the-minute manufacturing and sales techniques.

American-based importers like Naples-born Florindo Del Gaizo also adapted to the changing circumstances. A resident of New York since 1909, Del Gaizo served for many years as a counselor to the city's Italian chamber of commerce. The Italian chamber of commerce assisted importers based in the United States in various ways, such as providing them with the most recent prices so they could calculate import duties. Del Gaizo had a foot in both countries, as he continued to manage his tomato canning factory outside Naples, at San Giovanni a Teduccio, the source of his imports. His tomato preserves dominated the New York market. In the early 1930s, a Brooklyn woman, Tillie Lewis, noted the popularity of Italian canned whole tomatoes with her Italian customers, and in Del Gaizo she found an Italian partner to import the 'San Marzano' seeds. Together, they began to cultivate and can this variety in California, shipping the finished product to customers in the northeastern United States.

By 1937, twelve Italian American producers in California were making more canned whole tomatoes and tomato paste than were being imported from Italy. Although they may not have been tomatoes from Italy, they still could be labeled "Italian tomatoes," and consumers often did not notice the distinction. The American product labels also were in the colors of the Italian flag and used images, symbols, names, and personalities associated with Italy. Thus names like "Pomidoro Campania" and "Concentrato la Parmense," which were similar to, if not indistinguishable from, their Italian counterparts, were available on the same shop shelves (figure 23). Not

Figure 23 Canned sauce and paste, made from "Italian" tomatoes grown in California, are advertised in the New York newspaper *Il Progresso italo-americano*, September 7, 1930.

to be outdone, Italian exporters also tried to attract Italian immigrants. Giovanni Pandice, of San Giovanni a Teducccio, decorated his cans of "Conserve Aquila" with a flying eagle bearing a crate of tomato preserves between an erupting Mount Vesuvius and the Statue of Liberty.

Tomato paste, whether produced in the United States, imported from Italy, or made at home, was feeding a newfound taste as an ingredient in the development of specifically Italian American food habits and customs. According to Cinotto, the important role of food in family life and as a symbol of ethnic identity for successive generations of Italian Americans is the result of a complex historical process that unfolded in urban and industrial America. The result was a hybrid that in a sometimes paradoxical way looked back to Italy while taking advantage of what the new continent had to offer. This process began as an action/reaction by Italian immigrants to their new life and was continued, codified, and fixed by the following American-born generation. This tradition arose at a time when the host nation's food was changing as well, as the mechanization, standardization, and mass marketing that the immigrants associated with the food industry in the United States also were quite new.

The dietary culture of Italians in the Americas can be characterized by both continuity and change. The irony is that the immigrants themselves were often unaware of how much change was taking place as they strove to maintain continuity. When they left Italy, Italians brought with them as much as they could. In 1884, the Italian journalist and novelist Edmondo De Amicis recorded his impressions on board the steamship *Galileo*, bound from Genoa for Argentina. On approaching their destination, his fellow passengers started to worry about the customs duties on the goods they had brought with them:

And it was moving to hear what poor things they were worried about, presents for the most part for relations and friends in America: a bottle of special wine, a round of cheese, a salami, a kilogram

[2.2 pounds] of Genoese or Neapolitan pasta, a liter [1.06 quarts] of olive oil, a box of dried figs, even an apronful of beans—but from their own house, from that certain corner of the garden, which the relation or friend must surely remember.

As emigrants, they had left Italy because of "hunger," but as immigrants, nostalgia and longing quickly set in. This was not nostalgia for the "land of poverty," of course, but for the festive foods and the community to which they belonged. Consequently, they reproduced the food production and consumption patterns that were more dreamed of than actual in the world left behind. The "old country" became a mythologized place, which immigrant parents described to their children as a place where poverty and hunger coexisted with food that was good and natural and where they all ate together as a family.

Immigrants could afford to eat to the point of satiety. They could even waste, something previously unthinkable. As one of Leonard Covello's immigrant informants told him about life in New York's Italian Harlem: "In Italy we were poor, always on the verge of starvation. . . . Who could afford to eat spaghetti more than once a week?" he asked. "Don't you remember how our *paesani* here in America ate to their heart's delight till they were belching like pigs, and how they dumped mountains of uneaten food out of the windows?" This victory over hunger and a diet of subsistence was nevertheless a real conquest. Food and plenty were associated with well-being. In the words of Jerre Mangione, recalling the immigrant experience in the early twentieth century, "If you ate well, you felt well. And if you felt well, all was well with the world." Food remained at the center of the immigrant experience, conditioning the changing culture of the immigrants long after they had become established in their adopted country. In a short story by John Fante, first published in 1940 but referring to the early decades of the century, it is the character Dino Rossi, "who always spoke Italian," who most appreciates the importance of food, "his eyes closed reverently as he inhaled the wonderful aroma of Mamma's tomato sauce."

Emblematic of this newfound plenty and the centrality of food is the "Sunday dinner," the main meal of the week. This ritual brought together the extended family for a meal that was consumed slowly, beginning in midafternoon and extending into the late afternoon or evening. This kind of slow preparation and leisurely consumption was generally possible only on a Sunday, since the meal habits of Italian immigrants had to adapt to a busy working life. "Sunday dinner" excluded the outside "American" world, and it signified that the family was realizing the American dream.

At its edible heart was a special meat sauce for the pasta, or should I say "gravy," for it was much more than a simple condiment:

> She heard the women talk about making gravy, speaking to a husband or child, and Rosemary understood the significance of this. It meant, Don't you dare come home late. . . . It was a special summons, a call to family duty. The pleasure, yes, of familiar food, the whole history of food, the history of eating, the garlicky smack and tang. But there was also a duty, a requirement. The family requires the presence of every member tonight. Because the family was an art to these people and the dinner table was the place it found expression.

"These people" are, of course, the Italian Americans, described in one of the passages of Don DeLillo's novel *Underworld* (1998) set in a Little Italy in the Bronx. It conveys all that was bound up in the meal: "This food, this family meal, this meat sauce simmering in a big pot with sausage and spareribs and onions and garlic, this was their loyalty and bond and well-being."

The word "gravy" itself suggests a hybrid. The sauce is reminiscent of Pellegrino Artusi's first sauce for *maccheroni alla napoletana*, with its rich, slow-cooked combination of different meats and flavorings, held together with tomato paste. Gravy conveys plenty, and for the southern Italian peasant, only emigration made this possible. With its origins in the city of Naples, *maccheroni alla napoletana* evolved into a byword for Italian American cooking, expanding far beyond the immigrants with origins in Naples and its hinterland. But the word "gravy" conveyed something that "sauce" did not: perhaps its association with meat, indeed with the richest part of the meat, its juices, as well as the fact that it was thickened. How the word "gravy" came to be used in this way is something of a mystery. It is strange that this most English of culinary terms should indicate something so elemental to Italian American cuisine. Italian "gravy" is one of many terms in the Italian American lexicon that combine southern Italian dialect with American English. It must have begun as a family or neighborhood usage, an original linguistic improvisation by the generation born in the United States and more comfortable in English. Another possible origin is that the first packages of Italian macaroni included instructions—translated into English of a sort for the overseas market—on how to "dress" the freshly cooked pasta. The standard three suggestions were grated cheese, tomato sauce, or gravy. That is, "gravy" was the word chosen to translate the Italian *succo di carne*, which meant anything from meat juices to a ragout.

We can now find recipes for "classic" or "authentic Italian gravy," which followed the rediscovery of ethnic cuisine beginning in the 1970s. Their compilers often were unaware of the many levels of paradox present in that term. Many Italian Americans, especially in the northeastern United States, are familiar with the term, although not all use it, and it continues to engender much debate among Italian Americans.

The tradition of the "Sunday dinner" was constructed by Italian immigrants. It did not exist in Italy, certainly not for the peasants who emigrated. It was conditioned by financial resources, as well as the nature of the working week. Sunday dinner became the focus for the transmission (or, if you prefer, the inculcation) of cultural mores and aspirations from parents to children. The place of origin that parents described to their children on these occasions was not so much a real place as a place remembered, a place imagined. The immigrants gradually filled it with

Italian immigrants in the United States were too busy building a new life for themselves to write cookbooks, which makes it difficult to find early examples of Italian American recipes. For instance, Maria Gentile's *The Italian Cookbook* (1919) has several pasta recipes but nothing quite like "gravy." So I have chosen a more recent one from the many available today.

Sunday Gravy

For the sauce
2 tablespoons olive oil

1 pound meaty pork neck bones or spareribs

1 pound veal stew meat or 2 veal shoulder chops

1 pound Italian-style plain or fennel pork sausages

4 garlic cloves

¼ cup tomato paste

3 large cans of Italian peeled tomatoes

2 cups water

Salt and freshly ground pepper

For the meatballs
1 pound ground beef or a combination of beef and pork

½ cup of plain bread crumbs, preferably homemade

2 large eggs

1 teaspoon very finely minced garlic

½ cup freshly grated Pecorino Romano or Parmigiano-Reggiano

2 tablespoons finely chopped fresh flat-leaf parsley

1 teaspoon salt

Freshly ground pepper

2 tablespoons olive oil

To serve
1 pound pasta shells or rigatoni, cooked and still hot
Freshly grated Pecorino Romano or Parmigiano-Reggiano

To make the sauce, heat the oil in a large heavy pot over medium heat. Pat the pork dry and put the pieces in the pot. Cook, turning occasionally, for about 15 minutes, or until nicely browned on all sides. Transfer the pork to a plate. Brown the veal in the same way and add it to the plate.

(*continued*)

Place the sausages in the pot and brown on all sides. Set the sausages aside with the pork. Drain off most of the fat from the pot. Add the garlic and cook for about two minutes or until golden. Remove and discard the garlic. Stir in the tomato paste and cook for 1 minute.

With a food mill, puree the tomatoes, with their juice, into the pot. Or, for a chunkier sauce, just chop up the tomatoes and add them. And the water and salt and pepper to taste. Add the pork, veal, and sausages and basil and bring the sauce to a simmer. Partially cover the pot and cook over low heat, stirring occasionally, for 2 hours. If the sauce becomes too thick, add a little more water.

Meanwhile, make the meatballs. Combine all the ingredients except the oil in a large boil. Rinse your hands with cool water and lightly shape the mixture into 2-inch balls. Heat the oil in a large heavy skillet. Add the meatballs and brown them well on all sides. (They will finish cooking later.) Transfer the meatballs to a plate.

After two hours, add the meatballs and cook for 30 minutes or until the sauce is thick and the meats very tender.

To serve, remove the meats from the sauce and set aside. Toss the cooked pasta with the sauce. Sprinkle with cheese. Serve the meats as a second course, or reserve them for another day.

Allen Rucker and Michele Scicolone, *The Sopranos Family Cookbook* (New York: Warner Books, 2002), 14–15.

idealized constructions, which had a very real function for the immigrants: to interpret, explain, criticize, and even deny the New World present, to both themselves and their children.

Memory, as we know, can play tricks. Foods recollected as "traditional" by immigrants often turn out to be native to the adopted country. A Piedmontese peasant in Argentina singled out the eating of *asado* as a Piedmontese tradition. But *asado* is a mixed grill, native to the Pampas of Argentina. The hybridization of foodways by immigrants is an example of how culture is not fixed, something that can be cataloged. Rather, culture is a process, continually changing, and adapting to fit new circumstances. *Asado* becomes a sign of new dietary possibilities and opportunities for Italian immigrants. It is an element of change.

Just as often, change mingles with continuity, almost invisibly. The importation of tomato preserves from Italy would seem at first to be a good example of continuity. After all, tomato preserves already had an established place in the cuisine of many Italian regions. And yet, as we have seen, their manufacture in Italy was quite new, especially in the form of canned whole tomatoes. Furthermore, their importation was made possible by their continued low price, despite tariffs, and their practicality responded to life in America in the early twentieth century. Italian immigrants nonetheless remained suspicious of anything in cans, which they considered unnatural and "foreign," but they were prepared to make an exception for tomatoes, considered a familiar and traditional food.

Italian importers, producers, wholesalers, shopkeepers, and restaurateurs in the United States stressed the traditional and national nature of their products. They encouraged consumers to buy them out of ethnic solidarity, to spend their money within the Italian American community. In the 1920s and 1930s, New York City alone was home to some 800,000 Italian immigrants and their American-born children. They owned 10,000 grocery stores, 2,000 bakeries, 875 butcher shops, and 757 restaurants. They ensured that Italy could continue to export "traditional" products—pasta, tomato preserves, liqueurs—while having to import "modern" goods.

Somewhat paradoxically, Italian producers in the United States also stressed something very "American" about their goods: the modern manufacturing process. By marketing the application of modern industrial methods to traditional foods, producers linked the Italian producers to the industrialized and urbanized United States. By the time the mass immigration of Italians had begun, near the end of the nineteenth century, the American food industry was growing rapidly. Increasing mechanization and improved transportation links resulted in falling food prices. Distribution and corporate organization linked the country's many regions into a single, national marketplace. New food products became standardized. By 1900, the food industry accounted for one-fifth of the industrial production in the United States, but the food industry in Italy, by contrast, was still in its infancy.

The production and consumption of pasta are a case in point. At the same time that pasta was being imported from Italy, it also was being manufactured locally. The availability of durum semolina wheat meant that any house or grocery store could produce it. Visitors to New York's Little Italies were struck by the pasta being produced and hung up to dry from window frames, in basements, and on tenement roofs. In the words of M. Sherman, writing in 1906: "Macaroni is made in every block of the Italian neighborhoods of New York. In many streets you will find three or four little shops in one block of houses, with the macaroni drying in the doorways and windows. The front room is the shop, the family living in the middle and rear rooms."

Pasta became stereotypically "Italian" in the eyes of Americans. It did not matter that its regular consumption was a novelty for most Italian immigrants. But the stereotype had its origins in actual practice, as pasta soon achieved high status and became part of Italian Americans' ethnic identity, which posed a problem for those who origins were in non-pasta-consuming parts of Italy. Spaghetti with tomato sauce became the ultimate Italian American comfort food. In the Hollywood gangster film *Little Caesar* (1930), the outlaw Tony Passa, now abandoned by the protagonist (Rico Bandello) and awaiting his fatal vengeance, is comforted by his Italian mother. Dressed in black, wearing an apron, she says to him in broken English: "I have some spaghetti for you on the stove. . . . You feel better . . . eat somethin' . . . do you good." This was the first of many in films in Hollywood's "pasta paradigm" in which the tomato sauces prepared and consumed by gangsters echo the bloody acts they commit.

The packaging and branding of pasta began with Italian consumers in the United States. By the 1930s, they had developed an appetite for the American-made product, increasingly able to compete with Italian imports. They identified with brands like La Rosa. Indeed, V. La Rosa and Sons was typical of the "modern" American model of food industries in introducing the packaging of its product, in small cartons, at a time when most pasta was still sold loose. La Rosa and Sons also created a demand for its brand through newspaper and radio advertising. Founded in 1914, by the 1930s it employed three hundred workers at its Brooklyn factory, and by 1938 it was responsible for one-third of packaged pasta sales in Italian areas. In the following year, the Food, Drug, and Cosmetic Act established packaging requirements and standardized pasta types, which had the effect of concentrating pasta production in the hands of fewer, larger companies. Italian pasta exporters therefore had no choice but to comply with the new regulations.

> I am boarding with the Oronzo household and occasionally eat with the Liuzzi family. . . . I can barely stand their company. They are a bunch of ignorant *cafoni* [country hicks]. They slurp their soup, they shovel the pasta down their gullets like a bunch of pigs while smacking their lips and making their noodles whistle. . . . Liuzzi's uncle is the worst offender. His shirts are always spotted with tomato sauce. One of the girls put a large towel around his neck in place of a napkin to the hilarity of everyone.

This was how one Italian immigrant boarder, in Utica, New York, described a festive meal on one Sunday in 1907. From such humble beginnings, a new hybrid cuisine was born.

Inexpensive tomato preserves and inexpensive pasta were two key ingredients in the development of a unique hybrid, the Italian American restaurant. Although many Italian immigrant families were loathe to set foot in such a place (or at least admit to doing so), the dishes served did have an impact on domestic cooking. They also contributed to the derogatory depiction of Italians as "spaghetti-benders," an expression dating from the 1920s. The "red sauce joint"—with its dishes smothered in tomato sauce, its red-checked tablecloths, and its candles stuck in Chianti bottles—became stereotypical to the point of caricature.

The origin of the Italian American restaurant lies in the many boarding houses and basic restaurants that fed the many unattached men in the early days of mass immigration. Such simple institutions were part of the Italian immigrant network of food producers and suppliers discussed earlier. Single men, or *bordanti*, in the evolving Italian American language of the time, frequently boarded with other Italian families. In order to economize, they sometimes took their meals with them, too. They also ate in the basic "eateries" set up by other immigrants, like the one the aforementioned fictional Vittoria Ruiz lived in after her arrival in New York.

The restaurants were simple and undecorated, with sawdust on the floors and dirty windows, but the food they provided was economical and filling. Most common was the new Italian American hybrid of "spaghetti and meatballs," which was attractive to restaurateurs because it was both substantial and cheap to prepare. Meat was offered as well, which itself was something of a novelty, as we have seen. The dishes were standardized creations like "chicken cacciatore" and "veal scaloppini," although there were more spellings for the two dishes than recipes for them. But they looked vaguely Italianate on the menu: a hybrid language for a hybrid cuisine.

This was a great change from the Italian-run restaurants of New York at the start of the mass immigration. In the 1890s, these restaurants tended to have French names and French-style menus and to serve a largely Anglo clientele. Spaghetti would have been the only Italian presence on the menu, but it was listed as "Spaghetti Italienne." Rendered reassuringly French, the spaghetti would have differed in presentation, too, cooked until soft in a beef broth.

The Italian American restaurant was thus something new, a mixture of Italian and American like the hybrid languages evolving in Little Italies throughout the New World, which mixed standard Italian, Italian dialects, and the language of the host country. This is captured in the poem "Spaghetti House," by Naples native Vincenzo Campora, who settled in New York in 1914:

Tell me, sir, how would you like it, al dente?
Tomato sauce and parmesan cheese
all-Italian, you know, even the pan
the basil, the chef, the help . . .
But here, at my place, friend and servant
of every Italian and all good tradition
you will find the culinary art of the Homeland
where every flower smells and with what an aroma!

[Dica Signor: come li vuole, al dente?
"Tomato sauce" e "parmigiano cheese";
tutto Italiano, sa, pur la padella,
il basilico, il "chef," l'assistente . . .
Ma qui, da me, amico e servitore
d'ogni italiano e d'ogni buon "custume"
c'è l'arte culinaria del Paese
dove ogni fiore "smell" e di che odore!]

The irony is that by the time Campora wrote this poem, in 1935, few Italian immigrants and their families were eating in such places. They now had families, for whom meals were important domestic occasions. In New York's Greenwich Village, the area's "bohemians" started to frequent Italian American restaurants, in a spirit of cultural rebellion and adventure. For the price of an evening meal, they could have an "Italian experience." But it was Prohibition that brought this experience to millions of Americans. Wine could not be sold publicly, but it could be

made for "domestic consumption." As Giuseppe Prezzolini remarked in
Spaghetti Dinner (1955): "In more or less dimly lit basement speakeasies,
Americans could find wine, more or less palatable, made by an Italian,
more or less legally. And what would the Italian serve to go with his wine?
What else but his national dish? Thus proper spaghetti was informally
introduced to millions of Americans."

Although I would not go so far as to call spaghetti an Italian national
dish, the American public certainly thought of it as such, and they
expected "familiar" dishes. The absence of an Italian clientele furthered
the creation of a hybrid cuisine as it standardized the menu, eliminating
regional variations and specialties. If not strictly Italian, spaghetti and
meatballs was certainly an established Italian American dish by this point,
both in restaurants and at home. It had become emblematic. In Puzo's
Fortunate Pilgrim, Frank Corbo is invited to Sunday dinner, "the feast of
the week for Italian families. The spaghetti was coated with Lucia Santa's
finest tomato sauce, the meatballs were beautifully round and peppered
with garlic and fresh parsley."

Other hybrid dishes had their origins in the restaurant environment. *Cotoletta alla milanese*, a simple breaded veal cutlet transformed by
the addition of the ubiquitous tomato paste, became "veal parmigiano,"
a breaded veal cutlet topped with cheese and tomato sauce: the readily
available meat in the New World meeting the more traditional eggplant
and zucchini *alla parmigiana* recipes. Very close to this is Argentina's *milanesa napolitana*. Here, the *milanesa*, in the form of a "Milanese" veal cutlet,
moved to Naples in the form of the tomato paste covering it, along with
ham and cheese. In fact, the "Naples" portion of the dish may come as
much from the name of the Buenos Aires restaurant that first started
serving it in the 1940s, José Napoli's El Napolitano.

The most "American" twist in this tale is something else, however. If Italian immigrants in the United States were wary of food in cans, it was the
ultimate irony that their beloved spaghetti should end up in one. Canned
spaghetti in tomato sauce dates back to the early years of the twentieth

century, the idea of French immigrant Alphonse Biardot, founder of the Franco-American Company in 1887. The company already was famous for its tomato soup. In order to make its canned spaghetti attractive to non-Italians, Biardot downplayed its ethnic origins. The label thus read "Spaghetti à la Milanaise," "prepared according to the recipe of A. Biardot of Paris." Other companies—Heinz, Kraft, and Campbell's—quickly followed suit, with Campbell's bringing out Franco-American Spaghetti in 1921.

The most successful was a product launched by an Italian. Ettore Boiardi was a chef from Piacenza (Emilia Romagna), who opened his first restaurant in Cleveland in 1926. He decided to package his sauce together with grated cheese and dried spaghetti: a complete ready-to-make dish. His business was called the Chef Boiardi Food Products Company, but in order to make it more understandable to Americans, he changed the company's name to Chef Boyardee. "Everyone is proud of his own family name," Boiardi said, "but sacrifices are necessary for progress." The sacrifice coincided with the launch of a new product: canned spaghetti. His smiling face, complete with chef's cap and neckerchief, adorned cans of his "Spaghetti Boy-ar-dee." By the 1930s, it also was sold throughout the United States in the A&P supermarket chain, and by the end of the decade it had become an all-American food when it formed part of U.S. Army rations. This was quite a paradox at a time when Italy was about to become an enemy nation. Some Italian Americans were arrested and interned as "enemy aliens," but many more joined the U.S. Army. The sight of GIs opening cans of tomato spaghetti must have been a strange one to southern Italian peasants as the Allied forces made their way up the peninsula in the latter stages of World War II. The final irony is a small one. Boiardi died in 1985, in Parma, Ohio. One wonders what he made of all this, given the centuries-old rivalry between Parma and Piacenza in Italy.

It is easy to look down on such products, but it was a new way of eating food. After all, both spaghetti with tomato sauce and the invention of canning began about the same time, in the mid-nineteenth century, so why shouldn't they be united? It is just that we attribute different meanings, different values, and a different social status to *pasta al pomodoro* and canned spaghetti. From the point of view of the tomato itself, however, canned spaghetti represented another step in its use. Processed tomatoes were increasingly being used an ingredient in other, processed foods, even though the Italian path to modernization was quite a different one.

6

The Autarchical Tomato

The following sentimental poem was written by a middle-class house-wife of Turin in praise of a Cirio product. It was published in October 1927 to much fanfare in Italy's popular Sunday magazine *La Domenica del Corriere*:

> He, in a fury, huffs:
> "What a tasteless soup!"
> She sobs, downhearted . . .
> What a silly scene!
> The chambermaid enters,
> removes the soup tureen.
> A few minutes go by . . .
> The two are taciturn no more.
> The soup has been eaten,
> peace has been restored.
> She smiles: "What a fool
> to suffer so, such a waste . . .
> All it needed was Cirio tomato paste."

> [Lui furibondo, sbuffa
> "Che minestra scipita!"

Lei singhiozza avvilita . . .
Quale scenetta buffa!
Entra la cameriera,
riporta la zuppiera.
Passan pochi minuti . . .
I due non son più muti.
La minestra è mangiata,
la pace è ritornata.
Lei sorride: "Che scema,
soffrir tanto martirio . . .
Bastava un po' di Pomodoro Cirio."]

The poem was the culmination in the magazine of year-long campaign by the Cirio Company. Prizes of luxury holidays were offered for the submissions that, in the company's view, best expressed "satisfaction" and "pleasure in the goodness of the product, the delicacy of its aroma, and the attractive price." Tomatoes were now one of Cirio's leading products, and tomatoes now were a standard part of the Italian diet, in numerous guises. Indeed, tomatoes were part of the country's prosperity, reflected in Italy's new optimism and consumerism. For a time, Benito Mussolini and his Fascist regime channeled this mood into a flow of nationalist fervor and popular support. But within a decade, Italy was caught up in a downward socioeconomic spiral that brought the country to ruin and its people to the brink of starvation. But let us start on a happier note.

The mass migration of millions of Italians across the Atlantic had a positive effect on dietary practices in Italy in the form of remittances and return migration. The money painfully accumulated and regularly sent back home as remittances united the faraway Americas with even the most isolated Italian villages. The sums were an unheard-of influx of money for those who had stayed behind, and they provided a cushion against the effects of poor harvests. For the first time, these remittances also gave many Italians a chance to put aside money or goods. Storing food, or making *provviste*, which had been the privilege of the well-off, was now possible for peasants and urban laborers. As a result, the elaborate techniques of food preservation flourished as never before. According to the anthropologist Vito Teti, "dried, salted, smoked, acidic, and fermented foods [and] foods preserved in oil, salt, vinegar, sugar, water, lemon, and chilies tell of the effort and the habit to mitigate a direct dependency on seasonal availabilities and to do so in such a way that nothing was wasted."

Although mass migration broke the vicious cycle of climate-dearth-hunger that had kept the peasants tied to the land and to their landowners, return migration only made this worse. The returnees' expectations and aspirations had changed. For all the praise of "tradition" by Italian immigrants in the Americas, return migrants were not satisfied with the status quo they had left behind. These *Mericani* wanted more, and they sometimes could get it.

Until the mass migration ended by the 1930s, it was a two-way phenomenon. Migration was a way of life for the Italian poor, and rates of return and repeat migration were high, sometimes continuing for several generations within the same family. Between 1907 and 1911—peak years for Italians immigrating to New York—seventy-three of one hundred immigrants made a return trip. Some went back and forth between Italy and the Americas before opting for life on one or the other side of the ocean. The cash the returnees brought with them led to inflated salaries and expectations, with landowners having to pay their day laborers more or offer better working conditions. At the very least, therefore, the food got better, with bread, pasta, cured meats, cheese, and wine added to accompany the usual beans and greens. Needless to say, not everyone approved of the changes wrought by the *Mericani*. A study published in Florence in 1911, *La questione agraria e l'emigrazione in Calabria*, reported:

> [Emigration] has raised the level of nutrition in some places; some returnees buy meat, fish, etc. The class of small and large landowners rail against all this, shouting misfit and profligate. I

always remember the indignation of a wealthy landowner from the Nicastro area. "Formerly," he told me, "the peasant fed on corn bread and vetch; now he isn't afraid to buy luxury foodstuffs, fish, and meat!"

Where would things end? In fact, occupational mobility was limited, in both Calabria and elsewhere in Italy. At best, returnees managed to graduate from day laborers to lease holders or to the owners of small plots of land. Although there was little commercial or industrial development, the returnees did manage to improve their material conditions. They built better houses for themselves, as the remittances enabled improvements in basic material conditions, often in towns depleted by emigration. During this time, the changes in diet also gave rise to the recognizable regional cuisines of Italy, as people were able to spend more on food. This was a time of widespread economic prosperity in Italy. Commercial agriculture experienced a brief boom as output increased, new products were introduced, and consumption levels rose.

Migration had another effect as well. The presence of millions of consumers of Italian food outside Italy acted as a spur for the food industry within the country itself. This was especially true for foods regarded as "traditional," like pasta, processed tomatoes, olive oil, and hard cheeses. Indeed, foreign demand was crucial to turning certain foods into mass-produced goods, which was accompanied by increased consumer demand within Italy, and Italians in turn were able to take advantage of relatively low prices.

All this coincided with the ready availability of tomato products, especially tomato concentrate. In chapter 5, we looked at the Italian tomato from the outside in, as it were, from the point of the United States and Britain. What was the view of the tomato within Italy? The period from the 1880s until 1910 was a veritable boom for tomato processing. New companies sprang up and, just as quickly, disappeared, in a race to produce greater and greater quantities of concentrate. By 1912, unsold stocks filled factory shelves.

An example is the small town of Felino, outside Parma. Since the mid-nineteenth century, tomatoes had been grown there for the small-scale production of *conserva nera*, and in 1874, the first tomato-processing company was established in Felino by a group of local farmers. Despite the self-effacing name Società anonima di coltivatori per la preparazione delle conserve di pomodoro (Anonymous Company of Farmers for the Preparation of Tomato Conserves), the company successfully brought together farming and industrial expertise with an entrepreneurial spirit. Its aim was to open more workshops to prepare *conserva*, to improve the quality of seeds and seedlings available, and, of course, to increase retail sales. By the 1880s, the Anonymous Company was exporting to Britain and Argentina.

From the 1880s and the first mechanization of the preserving process, tomatoes slowly became a cash crop in the area, like sugar beets. In fact, the comparison with sugar beets is apt because the tomatoes cultivated were not eaten fresh; they were destined for processing into tomato paste. The cultivation of tomatoes grew alongside the region's production of milk and pork, which were transformed into the region's most famous products: cheese and cured meats. It is curious that during this boom period, of Felino's twenty-eight tomato-processing factories twelve also produced cured meats. In addition, three of the producers operated small restaurants from which they sold their produce. The tradition of producer-run restaurants in the region continues to this day.

Combining the curing of meats, a wintertime activity, with the processing of tomatoes, a late summertime activity, made good economic sense, as it also spread the financial risk. The region's large landowners were quick to understand the profits to be made in tomato cultivation and, especially, tomato processing, and the result was competition and overproduction. Only fifteen of Felino's twenty-eight processors, the larger ones, managed to stay in production for longer than five years. Many of the rest, the smaller ones, were lucky to survive even a season or two. By contrast, the companies founded after this early period tended to specialize in tomato products and also to stay in business longer, some into the 1960s.

The increasing mechanization and, with it, the higher cost of production were other factors in the concentration of processing into fewer, larger companies. At first, the industrial methods for making tomato paste resembled the domestic ones discussed in previous chapters. All that was needed was a copper kettle on the fire and some wooden boards for the *conserva nera*. Parma's industry then was given a boost in the early twentieth century with the importation of the first *boules* (vacuum boilers) from France. Used in the industrial production of sugar, the *boules* were large, bottle-shaped copper cauldrons, complete with a double bottom, revolving paddle, condenser, vacuum gauge, and tap. A single boiler could convert up to 475 gallons of tomato pulp into concentrate in around three hours. The boilers worked on the principle that a liquid under pressure boils at a much lower temperature than in the open air, which meant preserving more of the tomato's original color and flavor. In addition, the steam given off by one boiler could be used to heat the next one, and on down, thus saving on fuel. This enabled the efficient and standardized production of double and triple tomato concentrates. Such was the importance of the production of tomato paste in the Parma region's economy that companies like Zanichelli manufactured and sold not only the paste but also the advanced machinery needed to produce it.

The predominance of concentrates over the traditional *conserva nera* was ensured by the arrival of machine-made tin cans, which were not suitable for the drier *conserva nera*. Rather than relying on a tinsmith's painstakingly manufacturing cans by hand and then soldering them shut once they had been filled, machines now did the assembling. By the early twentieth century, they were made in standard sizes for tomato paste: 7 ounces, 1.1 pounds, and 2.2 pounds. The companies' labels often used similar colors, with red the most common, not surprisingly, plus green (to indicate freshness and the plant's leaves) and gold (to signify value and prestige). At a time when most Italians were still illiterate, especially the housewives most likely to buy the product, manufacturers tried to establish product recognition through the use of symbols. Parma's tomato processors favored eagles, lions, suns, and figures from mythology.

None of this would have been possible without the entrepreneurship of the region's landowning elites. As a class, they themselves were new on the scene: local businessmen and professionals who had invested in land, sold cheap by the impoverished aristocracy. Beginning in the 1890s, as tomato processing became increasingly industrialized and profitable, these landowning elites invested in it. Their investments became part of a transformation that paved the way for what is today Italy's world-famous "food valley," characterized by a close interaction between intensive agriculture and quality food processing. But because they were eager to farm their land as intensively as possible, the landowners turned away from the traditional tenant-farming agreements, which constituted the *mezzadria* system, in favor of salaried labor. Instead of a resident tenant who kept a percentage of the crop for himself, the farm laborers were paid only a daily wage and lacked any rights.

Antonio Bizzozero, who succeeded Carlo Rognoni as Parma's chief agronomist, warned that the shift would lead to social unrest. The region's large landowners were trying to set themselves up as a ruling class, controlling life in both town and country. By contrast, Bizzozero favored peasant cooperatives, the advancement of credit, and instruction in the most up-to-date farming techniques. His call for calm and mediation between the landowners and the peasants went largely ignored, however, and in 1907/1908, an anarcho-syndicalist strike paralyzed much of the region. It began with the salaried day laborers and spread to the towns, including Parma. Although the landowners were not afraid to use violence against the strikers, they returned to tenant contracts to farm their lands. From the peasant's point of view, the *mezzadria* system was a harsh one in many ways, but it was preferable to salaried labor, and at least they were able to reap some of the benefits from cultivating cash crops like tomatoes. This outcome also benefited the landowners, who renewed their attention to the processing industry, investing around 20 million *lire* by 1912.

Another feature of Parma's tomato-processing industry was the tensions between the producers and the suppliers, which also came to a head in 1912. Based on the high prices for tomato concentrate in the previous year, processors competed to contract for supplies of tomatoes, thereby pushing up the price. Bizzozero again advised restraint. But the price eventually rose so high that by the time the fruits had matured on the plants, the processors refused to buy them. Farmers were thus forced to let the tomatoes rot in the fields, suffering enormous losses, rather than accept a far lower price. Because the processors and producers failed to decide on a fixed price for the fresh tomatoes, as they had done for milk (to be turned into cheese), instability continued to plague Parma's tomato-processing industry for many years. At least both the region's producers and processors had other crops and other products to fall back on.

Despite the increasing mechanization of some phases of tomato production, most was still done by hand. However the actual labor was organized, whether by tenant farmers or hired farmhands, it was intensive and backbreaking, beginning with cultivation. The plants had to be trained on trellises, irrigated, fertilized, and eventually picked—all by hand (figure 24). The benefits of this method were the high production and relatively little waste. But it was feasible only because labor in Italy was cheap and readily available. Harvest was quite literally child's play. As Remigio Baldoni nonchalantly observed in *Il pomodoro industriale e da tavola* (1940), "Women and children are responsible for the harvest; the work is not heavy but requires much time."

After being harvested, the tomatoes were carted to the processing factories, which sometimes were simple structures located close to the fields. By contrast, the factories of Naples and Catania were substantial, well-lit buildings with stone walls, tile roofs, and cement floors. They were "of an ornamental construction rarely met with in American factories," in the words of Alexis Shriver, a "special agent" of the United States Department of Commerce. Once the tomatoes were brought to the factories, children and women took the tomatoes from their crates and sorted them by hand (figure 25). The tomatoes were then washed,

Figure 24 Labor-intensive tomatoes, grown on an Italian farm in Libya, 1930s. (Photograph courtesy of Paolo Cason)

Figure 25 The arrival of crated tomatoes at a factory, province of Bari. (From Remigio Baldoni, *Il pomodoro industriale e da tavola* [Rome: Ramo editoriale degli agricoltori, 1940]. Courtesy of the Biblioteca dei Georgofili, Florence)

placed into metal baskets, and briefly immersed in boiling water. Working on tables, women removed the skins from the still-hot tomatoes by pressing them between their fingers.

For canned whole tomatoes, the peeled fruits went straight into the awaiting cans and then were steam sealed. For tomato paste, after the tomatoes were peeled, chopped, and strained, the machines took over. The paste could be made from either whole tomatoes or the peelings and trimmings obtained from the whole, peeled tomatoes that had been canned. These two methods were seasonal, since both required fresh tomatoes. A third method for making paste used already manufactured and canned tomato pulp, a soupy, unrefined mixture known disparagingly as *minestrella*. Although it could be used all year-round, it was regarded as inferior in quality and was never very popular. Regardless of the method, once the paste was ready, it was ladled by hand into cans. The filled cans also were labeled and boxed by hand.

Boys made the wooden shipping crates. To work in the factory, children had to be at least twelve years old, and all children between twelve and fifteen had to have a valid *libretto* certifying their fitness to work. All they needed for their papers was to have completed primary school, to be healthy, and to have had their vaccinations. The same Italian law concerning the employment of children and women, dating from 1902, also stipulated certain minimal working conditions for women. Women constituted the bulk of the workforce for the manufacture of tomato paste (figure 26). Nursing mothers were allowed out of the factory for up to one hour a day to nurse their infants, and factories employing more than fifty women had to have a nursing room inside the factory. Although women and children were not permitted to work on Sundays, for men this day was optional.

According to historian Carol Helstosky, the twentieth century brought changes in Italy's consumption habits, changes not limited to the middle classes. In the working-class neighborhood of Testaccio in Rome, for instance, a study conducted in 1912 found that families there were able

Figure 26 Say cheese! A group photo of the workforce, cans held high, at Ceci, Gherini, and Company, Felino (Parma), early twentieth century. (From Pier Luigi Longarini, *Il passato . . . del pomodoro* [Parma: Silvia, 1998])

to eat a wide variety of foods, including meat. Based on an analysis of the spending habits of 1,300 families in the neighborhood, the study nevertheless reported with disappointment that "hundreds of families eat a *minestra* composed of lard, canned tomatoes, seasoning, and pasta for both lunch and dinner." What the study did not appreciate was just how much the diet and spending of Testaccio residents had changed, signaled by the consumption of pasta (partially taking the place of bread) and canned tomatoes.

The consumption of tomatoes per capita between 1900 and 1915 was almost double what it had been twenty years earlier. Consumption across the board increased as well, although tomatoes benefited more than any other product. Tomato paste even found its way into the diet of Tuscan tenant farmers, and boiled and peeled tomatoes are cited in the first Italian books of household management for women. Finally, they were a common ingredient in the recipes published in what is perhaps the earliest collection of Italian regional specialties, Vittorio Agnetti's *La nuova cucina delle specialità regionali* (1909), for a country "unified" barely fifty years earlier (figure 27).

The country had never been so prosperous. Wage hikes, increased agricultural production and imports, and the remittances of emigrants all provided a feeling of wealth. This well-being was concentrated in the industrial "triangle" of Milan-Turin-Genoa, which took off during the first decade of the twentieth century. A new urban working class emerged, along with a new middle class of service workers, state employees, and private-sector white-collar workers. Here, at least, the standard of living reached the European average, and Italian consumers could afford to spend money on nonessential foodstuffs.

Figure 27 United Italy: in this advertising poster for Spampinato canned goods, the tomato paste is Sicilian; the pasta (and its eater), Neapolitan; and the wine (Barbera), Piedmontese. The kitchen range indicates a well-off household. (From Pier Luigi Longarini, *Il passato . . . del pomodoro* [Parma: Silvia, 1998])

New products were developed to meet this demand. For example, dried pasta was marketed with the tomato already in it: *pasta al pomodoro* was 10 percent tomato paste, and spinach pasta was prepared in the same way. The great novelty, however, was the "medicated" pastas, which were fortified with health-bestowing ingredients, in an age when dietetics dominated the food industry throughout Europe. Last but not least was Fragrans, a mixture of dried and powdered herbs and vegetables, including tomatoes (figure 28). The product promised "the garden in your kitchen" and was the "perfect substitute" for fresh ingredients.

Italy's entry into World War I may have had an adverse effect on the country's exports, but it did little to dampen the country's prosperity. Wheat consumption increased to the point that Italy had to import one-third of its needs, since domestic production could not meet demand. The greater demand for wheat, however, came at the expense of other grains. When the war curtailed wheat imports from Russia and Romania, Italy bought wheat from the United States, Canada, and Argentina. (For Italians in the Americas willing to pay a premium for pasta imported from Italy, the irony was that it was probably made from American wheat.) Wheat imports increased an amazing 900 percent during the war, as the government also had to meet the demands of the military as the as civilian population.

Whereas earlier Italian governments had rarely intervened in food production and consumption, the wartime government kept a close eye on the food supply. By 1916, local authorities had fixed the prices of most grains, pasta, flour, and sugar, the result being a strengthening of buoyant prewar consumer trends.

The postwar governments, however, could not continue to satisfy consumers without destroying the stability of the state. Price controls were lifted immediately after World War I ended, and the result was spiraling inflation. For example, when the local authorities in Parma responded to the chaos by fixing the retail price of tomato paste, it was more than five times the price it had been just four years earlier. In addition, because of

Figure 28 Why use fresh when you can have powdered? Advertisement for Fragrans published in the newspaper *Corriere della Sera*, December 25–26, 1914, 4. The advertisement appeared throughout the winter months.

a shortage of tomatoes, paste manufacturers were forced to join a "mandatory consortium" that established the conditions and the prices of the tomatoes bought from farmers. That these restrictions applied only to Parma was said to have damaged the region's competitiveness with other tomato-producing areas.

Despite the worsening economic conditions, consumers' expectations remained high, leading to a wave of strikes, riots, and the sacking of shops. The two years after the war became known as the *biennio rosso*, the "red" referring not to tomatoes but to the radical nature of the protests. Even the looting of shops reflected the higher standard of living. Police reports from the time indicate that the looters did not target bakeries or flour mills, the staples of the past. Rather, they raided grocery, wine, meat, shoe, and fabric shops, making off with canned goods, soap, wine, cheese, fabrics, and leather goods.

Promising order and prosperity, Benito Mussolini's Fascist Party made the most of the chaos. As the party's policy began to develop, it set as its main goal the promotion of social order through the state organization of all aspects of society. Mussolini and his advisers favored a domestic solution to the chronic problem of food availability. If Italy could not produce enough food to feed its own citizens, if its resources and infrastructure were limited, the solution lay at home. Increasing imports only made the country more vulnerable. During the 1920s and 1930s, the Fascist Party's policy was codified, bringing about what it saw as a "revolution" in ideas about food and its consumption. The government strove to maintain public order, public health, and social equality. But more than anything else, it wanted to make Italy self-sufficient in food. If the aims were laudable, the approach was heavy-handed. Exports were limited; consumers were discouraged from buying non-Italian goods; and a campaign was launched to increase domestic wheat production.

The "battle for grain" is the best known of these policies. It was started in 1925, at a time when other European countries also were trying to secure their own stocks of wheat. In order to make Italy self-sufficient in wheat, more land was set aside for wheat cultivation, and new strains were developed. The campaign was a success in that it led to markedly higher yields, but it was not enough to make up for the volume of wheat previously imported. Mussolini's regime soon realized that it would have to reduce wheat consumption and conserve supplies. In 1929, four years after the "battle for grain" had begun, the government initiated a campaign pushing whole-wheat bread. It also established a national rice board, the Ente Nazionale Risi. The Italian production of rice exceeded demand, however, resulting in a surplus. Although rice was widely eaten in the areas of its cultivation—in a band stretching from the flatlands of Piedmont in the west to the Veneto in the east—it was not often eaten elsewhere in the country. To change this, special "rice trucks" toured the country, especially the south, distributing free samples and recipe booklets.

Fascist food policies worked together with a message praising sobriety and thrift, as well as the consumption of certain foods, as national virtues. Propagandists linked consumption patterns and favored foodstuffs to the national character, a message strangely reminiscent of the Renaissance notions discussed earlier, that particular foods were best suited to certain peoples.

The costs of the "battle for grain" outweighed the gains. The first cost was financial. In strictly economic terms, it would have been cheaper to import more wheat, as world wheat prices were actually declining at this time. But the campaign had another, unforeseen cost: its effects on other areas of agriculture. The irony is that just as the regime began to advocate more consumption of fruits and vegetables, its "battle for grain" made growing such products unprofitable. Rising prices forced down consumption, and with wheat being more profitable than feed crops, the availability of livestock declined. Meat became a precious commodity.

Nowhere was the clash between media message and actual production more evident than with the tomato. In 1931, the Touring Club Italiano, Italy's answer to the American Automobile Association, published a culinary guide to the country. Its *Guida gastronomica d'Italia* proudly proclaimed that the tomato "has gained the rights to citizenship in nearly all the regions of Italy." The white of the pasta, the red of the tomato sauce, and a sprinkling of parsley or basil recalled the colors of the Italian flag, the guide patriotically informed its readers. A few years later, in 1935, in the monthly newspaper of the Fascist Party's farmwives' section, sent to its 3 million members, proposed something similar. Italy's *massaie rurali* (rural farmwives) were told that they could make a "patriotic omelet" out of home-raised eggs, homegrown tomatoes, and garden greens. The recipe itself was not new, as the Jesuits in Rome had been eating something similar in the eighteenth century. But in the Fascist climate, the omelet's preparation and tricolor appearance testified to the women's support for the motherland even in their domestic activities.

It was a shame, then, that tomatoes just then were becoming harder to find. Canning factories were hit by Fascist legislation limiting the number of women in the workforce to less than 10 percent, supposedly to foster the cult of the *massaia*. On top of this, the "battle for grain" made tomato cultivation much less attractive, and the start of the worldwide Great Depression of the 1930s reduced the demand for tomato exports. In 1928/1929, Italy was still cultivating 152,000 acres of tomatoes, producing 204,480 tons of preserved tomatoes. Just four years later, in 1932/1933, this area had been reduced by one-third, to 100,000 acres, and production was down to less than half, at 93,818 tons. Tomato exports fell by one-third. Processors went out of business. In the tomato-producing area around Salerno, home to the 'San Marzano' tomato, the number of pre-serving companies was almost halved, from 105 in 1927 to 60 in 1932. In the province of Parma, tomato cultivation declined while wheat, maize, and sugar-beet cultivation increased. Not surprisingly, Italians ate fewer tomatoes, with per capita consumption falling from 46.2 to 35.2 pounds during the 1920s and 1930s.

Notwithstanding this decline, Italy did show some signs of economic prosperity and modernization during the period. Along with the up-scale department store Rinascente, which had changed its name at the behest of the celebrity-playwright Gabriele D'Annunzio in 1917, two lower-scale stores opened: UPIM, in 1928, and Standa, two years later. Items of everyday consumption, from pasta to canned goods, were branded, packaged, and advertised. Although Italy's Fascists may have been worried about the rise of mass consumption, they did not want to stifle this development, seeing consumption and consumer goods as important to the construction of a Fascist society. This may seem like a paradox, but policymakers believed that standardizing the kinds of goods used and how they were used would contribute to the realization of a common national culture and generate a sense of belonging to this new, national collectivity.

Fascist food policy was at least partially successful during the 1920s. Partly because of the relative stability of the economy during these years, consumers did not suffer. The middle classes adapted in what has been called a "cooking of consent," making the advice and propaganda of Mussolini's regime their own. Books and magazines dedicated to food and the home proliferated as never before. If the consumption of meat had to be reduced, a virtue was made of the increased consumption of fruits and vegetables. Soups, stews, and pasta dishes all were advocated as healthy and economical. The high-brow gastronomic review *La cucina italiana*, which first appeared in December 1929, tried to strike a compromise between the regime's call for economy and its praise of the able housewife, on the one hand, and the appearance of the first labor-saving devices and products, on the other. Mayonnaise could be either bought or prepared at home, the magazine suggested, and canned tomato products provided a worthy substitute for home-made ones.

The culinary literature also explored regional diversity. Under Fascism, regionalism was seen as a positive attribute, a feature of Italian culture and identity—*Italianità*—which could counter the creeping and pernicious Anglo-American influences.

It is easy to assume that all these restrictions and propaganda campaigns stifled the culture of food in Italy, but nothing could be further from the truth. The Italian media were full of animated discussions and debates about food, and nothing raised tempers more than Filippo Tommaso Marinetti's proposal in *La cucina italiana* to abolish pasta, of all things.

The review was less than six months old when a member of its "tasting committee," a fellow of the Academy of Italy and a supporter of Mussolini, first mentioned something called "Futurist cuisine" in an editorial in *La cucina italiana* in May 1930. Marinetti was one of the founders of the artistic movement known as Futurism, begun two decades earlier and dedicated to a quest for "speed" and "modernity." In 1930, Marinetti announced on Italian radio that this new Futurist cuisine would have "as one of its principles the abolition of pasta." With the Christmas holiday season in full swing, on December 28, the full "Manifesto of Futurist Cuisine" appeared in the Turin newspaper *La Gazzetta del Popolo.* It was reprinted in the January 1931 issue of *La cucina italiana,* accompanied by a history of pasta by Paolo Buzzi and a spirited defense by pasta producer Vincenzo Agnesi. It was just the sort of subject that appealed to the new publication, which aimed to be intellectual and innovative and included numerous contributions by women authors.

What was all the fuss about? Marinetti had already ruffled feathers in his radio talk by defining pasta as "demoded." In Futurism's search for a cuisine that was "ethereal and rapid," pasta "weighs down, uglifies, deceives on its nutritional capacity, renders skeptical, slow, and pessimistic." Moreover, alluding to the regime's stated desire to conserve grain supplies, "it is patriotic to favor [Italian] rice in its place." The "biquotidian [twice-daily] pasta pyramid" was an "absurd Italian gastronomic religion." Look what it has done to the Neapolitans, Marinetti warned:

> Pasta contrasts with the lively spirit and impassioned generous intuitive soul of the Neapolitans. They have been heroic fighters, inspired artists, moving orators, acute advocates, resolute agriculturalists, all despite the voluminous daily *pastasciutta.* As a result of eating it, they develop the typical ironic and sentimental skepticism that so often stifles their enthusiasm.

Why do Italians continue to oppose "the immense network of long- and shortwaves that the Italian genius has launched over oceans and continents" with pasta's "heavy block"? Marinetti asked. Rather, Italians should be striving for "an agility of bodies" better suited to "the very light aluminum trains that will replace the present heavy iron-wood-steel trains."

Without pasta, what would happen to tomatoes? If pasta was a victim of its stodgy predictability in a cuisine that proposed "the absolute originality of dishes," the tomato was perceived as light and adaptable. Marinetti welcomed it as an ingredient in Futurist cuisine. The tomato could be found in several "formulas," the term Marinetti preferred to "recipe." The formula for "Country Crime"—*fattaccio campestre* in Italian, for the names were as imaginative as the compositions—calls for eggplants cooked in tomato sauce, stuffed with anchovies, and served on a bed of spinach and lentil pulp. A "Painting-Sculpture Summer Lunch" proposes a light repast for artists consisting of polenta, "good tomato sauce," salad leaves, apple slices, radishes, and white roses—inclusive of branches and thorns. The artists were to help themselves with their hands, eating what they were able to pick up. Marinetti's formula for "Alaska Salmon in the Sun's Rays" was not as risky for diners.

For those restaurant diners who insisted on spaghetti with tomato sauce, Marinetti proposed an alternative. The waiter was to inform the diner that the "silent, long, archaeological worms" had been eliminated from the menu, "white worms that you must not introduce into your body if you don't want to render it as closed, dark, and immobile as a museum." The waiter instead would serve the diner the "Modernizing Dish," made of "boiled rice, fried in butter and rolled inside small spheres of raw lettuce, sprinkled with grappa, and served on a pulp of fresh tomatoes and boiled potatoes." In this tempting formula, Marinetti uses the Italian *poltiglia* (pulp or mush) for the French *purée*. The "Manifesto," in line with Fascist ideology, opposed the use of foreign words, so it offered alternatives, some of which stuck, with others disappearing into deserving oblivion. The restaurant's *lista* (menu), where you might see a *pasticcio* (flan) and a *zuppa di pesce* (bouillabaisse), continued to be used. But the *pranzoalsole* (picnic), at which you might serve a *poltiglia* (purée),

Alaska Salmon in the Sun's Rays

To prepare *Alaska salmon in the sun's rays with Mars sauce*, take a nice Alaska salmon, slice it, and grill it with pepper, salt, and good olive oil until golden. Add tomatoes that you have previously sliced in half and grilled with parsley and garlic.

When it is ready to serve, place anchovy fillets crisscrossed in a checkerboard pattern on the salmon slices. Place a round slice of lemon with capers on each slice. The sauce is made up of anchovies, hard-boiled egg yolks, basil, olive oil, and a small glass of "Aurum"* liqueur, all of which have been strained.

Filippo Tommaso Marinetti and Fillìa [Luigi Colombo], *La cucina futurista* (1932; repr., Milan: Marinotti, 1998), 23.

* "Aurum," a sweet liqueur made from brandy and oranges, was all the rage during this period and was made in Pescara, the hometown of the poet Gabriele D'Annunzio (who also was responsible for its name).

or the *quisibeve* (bar), where you might drink a *polibibita* (cocktail), did not catch on.

The proposal to abolish pasta brought Neapolitans out onto the streets in protest. The city's mayor affirmed that "the angels in Paradise eat nothing but vermicelli with tomato sauce." A contributor to the July 26, 1936, issue of *Il Mattino*, Naples's main newspaper, reminded Marinetti of pasta's spread worldwide and its popularity in Europe's gastronomic capital, Paris. The article, "The Symbolic Macaroni," notes that "our most Italian friend F. T. Marinetti did not foresee the political function of spaghetti; otherwise he would never have condemned the patriotic *pastasciutta*." Indeed, even some of the most supportive Fascists criticized the "Manifesto." When Marinetti asked what kinds of attitudes and foodstuffs were best suited to "an imperial culinary art," the Fascist general Emilio De Bono—a veteran of the 1922 March on Rome and the marshal of Italy—replied with his version of the Holy Trinity: "I believe that only *pastasciutta*, risotto, and polenta are and will be, world without end" (*La Scena Illustrata*, June 1938).

At least one Neapolitan was in favor of Marinetti's proposals. V. G. Pennino, the lead writer at *La Gazzetta del Popolo*, suggested that along with pasta, wasteful white bread also should be eliminated from the national diet, in favor of "aromatic and substantial" whole-wheat bread. Marinetti probably would not have approved of the adjective "substantial," but he might have agreed with Pennino's praise of vegetable consumption, especially when couched in such lively, nationalistic terms. "Every people must have its own diet," Pennino wrote, in a tone strangely reminiscent of Renaissance physicians. The diet "of the Italian people must be based on the products of this hot, restless, volcanic land." Pennino recommended a diet consisting of three-quarters "marvelous vegetable products, for which we are envied by all the world," and one-quarter meat products.

Pennino was not the first Italian to write in praise of vegetables; the tradition goes back to the seventeenth century, as we saw in chapter 2. Tomatoes became increasingly fashionable with the rise of a new vegetarian sensibility, brought to Italy by Swiss and French visitors. This called for fresh ingredients, for the chemical contents of vegetables, and, beginning in the 1930s, for vitamins. This may explain why the Milan-based pharmaceutical firm Carlo Erba produced its own "triple concentrate" tomato paste.

The new taste for vegetables also is evident in *Cucina vegetariana* (1931), the cookbook published by the Sicilian aristocrat Enrico Alliata di Salaparuta. Alliata, scion of a famous wine-producing family, was an expert in French cooking and regional foods, a follower of Indian philosophy, and a firm believer in the health benefits of raw foods. He believed that fresh vegetables, rich in salts and vitamin C, had an important role in stimulating the appetite. Alliata's "manual of naturist gastrosophy" accordingly contains recipes for tomatoes served on toasted bread (*crostini*), in sandwiches, and in salads, all served raw with herbs and onions.

Alliata's fellow Sicilians subsisted on a diet that was likewise strong in raw vegetables, although more out of necessity than out of choice. Their diet was not unlike that of the inhabitants of Basilicata, as described by Carlo Levi. The poor ate "plain bread the whole year-round, spiced occasionally with a carefully crushed raw tomato." Alliata might have approved of this, but he might not have been so enthusiastic about the garlic and oil or the chilies also used to season bread. Making a virtue of necessity, the inhabitants of Puglia praised simplicity with the proverb "He who eats bread and tomato to the doctor will never go" (*Ce mange paene e pemedaore nan ve me' o dattaore*).

Despite the relative abundance of food during this period, there still were "two Italies." Southern Italians ate 50 percent more bread and more vegetables and legumes than northern Italians did, although not by choice. A survey of 744 working-class families conducted in 1937 found that the average northern Italian family consumed twice as much meat and animal fat as did its southern Italian counterpart. It also ate more than twice as much sugar, twice as many dairy products, and twice as many eggs. Normally, government statisticians preferred to aggregate the figures for the nation as a whole, thereby hiding the reality of southern poverty.

Alliata's serious and somewhat pretentious manual would probably have gone over the heads of most of his fellow Sicilians. More down to earth was another campaign promoting vegetable consumption, this one on behalf of the Cirio Company. The advertising campaign was masterminded by the American-trained Luigi Dalmonte in 1929. Dalmonte conceived of Cirio products in terms of self-presentation, along American lines. Advertisements showed how Cirio canned vegetables would save time in the household, create healthy children ("Moms, give your children Cirio's 'A-B-C' Tomato Juice, the perfect vitamin-based product!"), cure intestinal complaints, and provide a healthy way to lose weight, not to mention enabling housewives to acquire the envious admiration of their neighbors. Dalmonte's advertisements for Cirio sometimes promoted the Italian-ness of their products, but otherwise nationalist arguments were conspicuous by their absence. This changed, however, after 1936, even if the advertising strategies remained American in inspiration, and the contradictions are apparent in the strategies and content of Cirio's advertising campaigns, like the one at the beginning of this chapter.

In the 1920s and 1930s, the number of advertising messages aimed at housewives rose. The ideal housewife of the Fascist period was a consumer as well as an administrator, responsible for household purchases and domestic finances. When the regime's propaganda campaign for national self-sufficiency got under way in the late 1930s, advertisers promoted autarchical products and advice. They thus promoted Lanital instead of wool, *carcadè* instead of tea, and lignite instead of coal. But this did not mean that Italian housewives had lost their inventiveness and verve. At a time when food preparation was becoming even more a matter not just of frugality but of survival, Italian housewives still wanted to make a positive impression. Cirio's 1935 recipe collection, *Nuovi orizzonti per la vostra mensa*, contained three hundred recipes chosen from those sent in by Italian housewives. Its 1936 contest for the best use of canned tomatoes attracted more than three thousand contestants. Tomatoes were stewed in Marsala wine, in a cream sauce, or as an ingredient in lasagne. One contestant even proposed a concoction called "Pretend Birds in a Nest," with potatoes for the nest, veal and bacon birds, and tomato paste as a glue. At a time when few Italian households owned books, Cirio was easily the country's best-known publisher.

Following the American lead, Cirio also sponsored programs, like evening concerts, on Italian radio. It also launched its own brand of ketchup. A 1926 advertisement, designed by Dalmonte, boasted that "even in the most fashionable restaurants, where people dance the shimmy . . . you will find the piquant sauce Cirio Tomato ketchup." Fascism was not too taken with American dance trends, much less the use of foreign words. So Cirio launched a competition to come up with a more "Italic"-sounding name for its ketchup. The two finalists were Rubra and Vesuvio. Rubra won. Aside from being the Latin word for "red," it also recalled a "red sauce" (*salsa rubra*) common in Piedmontese cookery, used to season boiled meats, in which tomatoes provide the color. The product image changed, too.

The jazz, fancy restaurant, and elegantly dressed diners were replaced by an aproned housewife pouring "Cirio Rubra salsa piccante" over a chicken.

Cirio was not the only company canning tomatoes and condiments. In this time of product innovation and the first convenience foods, the Althea Company of San Lazzaro (Parma) introduced a prepared sauce called Sugoro in 1932. The name brought together *sugo* (sauce) with the "golden" of *pomodoro*. Althea also used avant-garde graphic designers in its advertising campaigns (figure 29). But it was Cirio that managed to keep itself most in the public eye. Its favored strategy, familiar to North American consumers, was the cookbook, in which the product had pride of place. In 1939, the Cirio Company published yet another book of three hundred recipes for its canned tomato products, which promised "fresh tomato in every season" (figure 30). Cirio's "Super Pomidoro Pelati," were "equal to fresh," in addition to being more convenient to use in cooking. Tomatoes went with everything—meat, fish, vegetables, pasta, and rice—and could be used in every course, from antipasto to dessert. But more than that, the tomato was "of itself the healthiest food, the most 'solar,' rich in vitamins, salts, of curative elements for uric acid and rheumatisms." Cirio's version of the tomato was not just "solar" but also southern. Its cookbook had recipes for Bari-style spaghetti, L'Aquila-style mushrooms, Sicilian-style swordfish, as well as southern recipes for vermicelli, vegetable soup, and pizza. But Cirio's canned tomatoes did not have to be confined to traditional dishes. The curious recipe for an antipasto called "Tomatoes on the Lawn" seems to owe something to Futurism, and "Sweet Pie" is a new twist on an old favorite, *crostata*, in which jam is used instead of fresh fruit.

Cirio had Lidia Morelli write the preface to its 1939 recipe collection, which was shrewd. Morelli was a an early-twentieth-century version of today's "kitchen goddess," having begun her career with an "encyclopedia of domestic life" in 1905: *Dalla cucina al salotto.* Morelli coordinated several recipe collections on behalf of the Cirio Company, including *Nuovi orizzonti per la vostra mensa.* Not only was she well known and well

Figure 29 Althea's ready-made Sugoro, which promised to season any food without having to be cooked, is attractively advertised in this unsigned poster, 1937. (From Pier Luigi Longarini, *Il passato . . . del pomodoro* [Parma: Silvia, 1998]. © Althea S.p.A., Parma)

Figure 30 The cover of Cirio's 1939 recipe collection, *Pomodoro fresco a ogni stagione*, gives the 'San Marzano' tomato pride of place.

Pomidoro sul prato (Tomatoes on the Lawn)

Ingredients: A 14.5-oz. can of "Super Pomidoro Pelati," 1 cooking apple, 1 tablespoon bread crumbs, salt and pepper, 1 Belgian endive or radicchio, a few sprigs aromatic herbs (marjoram, chervil, thyme, mint), grated lemon rind, [olive] oil, and lemon juice.

Slice the endive into thin strips. Finely chop the herbs and the lemon rind. Mix the herbs with the endive, and season like a salad with oil, salt, pepper, and lemon juice. Separately, grate the apple, and add 1 tablespoon bread crumbs and a pinch of salt. Mix in the "Super Pomidoro Pelati" to make a rather thick paste, and place in little spoonfuls on top of the previously prepared greens.

Crostata dolce (Sweet Pie)

Ingredients: For the pastry: 2.5 cups flour, ½ cup butter, grated lemon rind, small cup white wine, and a pinch of salt. For the jam: 21-oz. can "Super Pomidoro Pelati," ¼ cup butter, pinch of salt, 1½ cups sugar, 7 ounces amaretti [almond macaroons], and 3 eggs.

Make the pastry, stretching and rolling it 3 or 4 times, but without working it too much. Form it into a ball and let it rest. Cook the "Super Pomidoro Pelati" with the butter and sugar, stirring often, until it reduces into the consistency of jam. Crush the amaretti and mix them into the tomato mixture, then cook for another 2 to 3 minutes. Remove from the stove and mix in the eggs. Roll out the pastry dough, and, putting one-quarter aside, line the bottom and sides of a buttered pie pan. Pour in the jam, and use the remaining pastry to make strips for a lattice top. Sprinkle sugar over the top, and bake for half an hour.

Cirio, *Pomodoro fresco a ogni stagione* (San Giovanni a Teduccio: Cirio, 1939), 10, 87.

respected, but she was an enthusiastic supporter of the Fascist regime, which was important, because by the mid-1930s the climate in the country was changing quickly.

In 1935, Morelli published an updated and "modernized" version of her domestic manual, and she also rushed out an eighty-one-page booklet with the strident title *Le massaie contro le sanzioni* (Housewives Against Sanctions). When an empire-hungry Italy invaded Ethiopia earlier that year, the world community, in the form of the League of Nations (forerunner to the United Nations), slapped trade sanctions on Italy. Although the sanctions remained in place for only eight months, they were a forerunner of what was to come. Morelli's booklet offered advice on how to economize—how to make soups from the skins of legumes, coffee from toasted grape seeds, and comforters stuffed with old newspapers. It concludes with the Fascist motto, but adding Morelli's own resolutely cheerful touch: "Believe, resist, fight . . . but sing, too" (*Credere, resistere, combattere . . . ma anche cantare*).

Italy's African colonies—Libya, Somalia, and Ethiopia—were part of Mussolini's drive to turn the country into a world power. The regime was trying to counter the effects of decades of emigration, so instead of leaving for "America"—difficult in any case, since the United States imposed immigration quotas in 1924—Italians were to be sent to the country's own rural areas, its drained marshlands, and its African colonies. Hence the Fascist slogan: "Today, the real America is Libya." By 1940, some 300,000 Italian settlers, mostly peasants, were living in Italy's African colonies. This may sound like a lot, but it was much smaller than the number of Italians living in either New York City or Buenos Aires, and only slightly higher than the number of Italians who had left Italy each year during the 1920s.

The Fascist regime hoped that the African colonies would resolve Italy's chronic food shortages. In Italy, the political significance of food shifted. Now the face of the regime was less one of benevolence and more one of determination, before an obedient citizenry. Mussolini proclaimed *autarchia* (autarchy) in 1936; it was a catchword for the regime's policy toward self-sufficiency, at the heart of a massive propaganda drive. All aspects of the economy were strictly controlled. Even the land area devoted to tomato cultivation was decreed by the state. Thus as of 1938, farmers wishing to grow tomatoes required ministerial approval (law 736, regulating the cultivation of tomatoes for industry, in case you are interested).

Central to the policy of autarchy were Italy's African colonies. Elisabetta Randi's "autarchical" cookbook, *La cucina autarchica* (1942), optimistically described the introduction of tomato cultivation in Libya. But Libya remains a net importer of tomato products, just as it was in the late 1930s, despite some initial successes cultivating tomatoes on sandy soil (figure 31). In fact, far from being producers of food, Italy's African colonies absorbed one-quarter of Italy's exports after 1936, which was small recompense for the minimal quantities of bananas, peanuts, and sesame seeds that Italy imported from them.

Figure 3 1 Tomato cultivation by Italian colonists (with some local help) on Libya's sandy soils, around 1934. (Photograph courtesy of Paolo Cason)

Figure 32 The "little black face" began to appear on labels, such as this for canned tomatoes, and in advertisements following the invasion of Ethiopia in 1935. "Faccetta nera" was also the title of a Fascist marching song, with music by Mario Ruccione and lyrics by Giuseppe Micheli. (From Pier Luigi Longarini, *Il passato . . . del pomodoro* [Parma: Silvia, 1998])

Nonetheless, Italy's colonial efforts were broadly supported by the country. The years after the invasion of Ethiopia and before the start of World War II may have seen economic decline in Italy, but the country was optimistic about its future as a major actor on the world stage. Exotic Ethiopian faces appeared in popular songs and advertisements (figure 32). A "countersanctions song" to be sung in schools tells us something of the tenor of the times:

The potato and tomato
Have so many vitamins!
We happily sing in unison:
"England will never mine us."

Bread and salt, rice and pasta
Vegetables and fruit, cheese and wine;
Is all this country needs
For our simple morsel.

[La patata e il pomodoro
hanno tanta vitamina!
Noi cantiamo lieti in coro
"L'Inghilterra non ci mina."
Pane e sale, riso e pasta
erba e frutta, cacio e vino;
Ecco quanto in patria pasta
per il sobrio bocconcino.]

La cucina italiana, which after 1934 had become a thirty-two-page magazine for women, with articles on fashion and beauty, changed from being the *Review of Gastronomy for Families and Gourmands* to the *Review for Italian Families and Women: For the Resistance and Conquest of Economic Independence.*

Advertisers likewise followed suit. Only a few years earlier, in 1933, the manufacturer of Marca Bianca canned tomato paste had assured cooks that it was a false economy to make tomato paste at home. Either the paste ferments or the jars do not seal properly, to say nothing of the time spent and the gas used. Moreover, whereas the housewife has to boil the tomatoes at 100 degrees, causing them to lose color, flavor, and vitamins, the manufacturer of Marca Bianca could use "modern industrial procedures," which required lower temperatures and so "maintained the properties of the fresh fruit." Such a message was not in keeping with the new policy of autarchy. So by 1940, the company was boasting that the tomato seeds, formerly a useless by-product of the manufacture of tomato paste, were now being turned into an edible oil. Tomato-seed oil could "compete with all the cotton oils that come from the Americas." The company sent its tomato peels to Bologna, where it was hoped they could be transformed into a synthetic gum used to line the inside of tin cans. The rest of the waste was turned into animal feed. None of this was new; it all had been proposed by Renato Rovetta in his manual *Il pomodoro* (1914). But the changing economic and political times had made it a useful advertising message.

Meanwhile, Italians were encouraged to "imitate the Duce." The thirty-page pamphlet *Sapersi nutrire* (How to Nourish Yourself), by Cesare Alimenti, was printed by the Fascist Party's propaganda office. It contained advice, dietary norms, and health warnings geared toward a reduction in consumption. Readers were invited to imitate that person "who more than any other Italian develops an enormous amount of energy in his intense working day, but who nevertheless eats with a peasant, even Franciscan, simplicity."

Mussolini's abstemious habits, combined with his fitness, agility, and sharp wit, were legendary—quite literally, as they were part of Fascist propaganda. So in regard to his dietary self-discipline at least, the message was largely accurate. Mussolini's own simple social origins, together with a long-standing stomach ulcer, meant that he was never very interested in eating. Although he rarely ate meat, he did consume large quantities of milk, herbal teas, and fruit juices. While the rest of his family sat down to a first course of "Tagliatelle in Tomato Sauce with Shallots," a recipe of Mussolini's wife, Rachele, Mussolini would have noodles in broth, apparently his favorite food.

While Italy was under sanctions, Nazi Germany began exporting much-needed coal to the country, which led to a closer alliance between the two nations, culminating with the signing of the "Pact of Steel" in May 1939. The word "steel" gave a resounding ring to this alliance, but it might as well have been called the "tomato pact" (figure 33). By this time, Italy was exporting virtually all (90%) of its fresh tomato crop to Germany. German consumers were particularly fond of the large round, smooth varieties. That same year, Nazi officials toured Italy, evaluating Italian food production for other possible exports to Germany. Italy

Tagliatelle in Tomato Sauce with Shallots

Ingredients for six people:
4 cups flour
5 eggs
Grated Parmesan cheese (1 tablespoon)
Nutmeg
4 shallots, finely chopped
2 garlic cloves
6 peeled tomatoes
Olive oil, salt, and pepper

Knead together the flour, eggs, grated Parmesan cheese, and a pinch of nutmeg, and let it rest for about an hour. Roll the dough into a thin sheet for the tagliatelle. Prepare the sauce in a pan by lightly sautéing the finely chopped shallots and garlic cloves in olive oil. Add the tomatoes, skins and seeds removed, salt and pepper to taste, and simmer over a low flame. Cook the tagliatelle in plenty of salted water, drain when al dente, and pour into the pan, mixing well to cover in the shallot sauce. Serve immediately.

Maria Scicolone, *A tavola con il duce: Ricette e racconti inediti di casa Mussolini* (Rome: Gremese, 2004), 20–21.

also was exporting its laborers, and by 1942 some 300,000 Italians were working in Germany, many in the mines, extracting the coal that Italy got in return.

Italy entered World War II in 1940, as aggressive as it was underprepared. The country's food policy was in disarray, and because of its alliance with Germany, Italy's options for imports were limited, and its wheat production fell drastically. As the Albanian and Greek campaigns dragged on, and barely six months after Italy's entry into the war, rationing was introduced. It started with coffee, sugar, olive oil, lard, rice, flour, and pasta, and in October 1941 bread was added to the list of rationed foods. To add insult to injury, the rationed bread and pasta also tasted

Figure 33 The "tomato pact": fresh tomatoes being sorted and boxed for export to Germany, Italy's main market. (From Remigio Baldoni, *Il pomodoro industriale e da tavola* [Rome: Ramo editoriale degli agricoltori, 1940]. Courtesy of the Biblioteca dei Georgofili, Florence)

terrible, perhaps doing more than years of government campaigns to persuade southern Italians to eat rice—assuming they could find any.

The rationing of staples took account of differences in regional food patterns. Everywhere the ration was 4.4 pounds, but policymakers tried to factor in regional habits and preferences. In northern Italy (from the region of Emilia northward), the ration card permitted the purchase of 2.2 pounds of rice per person a month, 21 ounces of pasta, and 14 ounces of polenta. In central Italy (the regions of Liguria, Tuscany, Umbria, and Marche), it allowed 2.2 pounds of pasta, 21 ounces of rice, and 14 ounces of polenta. In the south (from the Lazio region southward), the ration was 3.9 pounds of pasta and 7 ounces of rice (but no polenta). All this planning was academic, though, as prices skyrocketed because of the poor implementation of price controls.

The breakdown in the food supply was soon evident to all, losing the people's trust in the regime, despite claims by the Propaganda Office of the Fascist Party that "no Italian has died from hunger, that no Italian will die of hunger, that in Italy one eats better than in any other part of Europe." An idle boast it might have been, but with much of Europe in the grip of rationing, if not outright want, it was not much to brag about. In Britain, the Ministry of Food printed a series of leaflets on how to cope with rationing and shortages. One of them described how to bottle tomatoes at home, appropriate given that Britain was the main importer of canned Italian tomatoes, now in short supply.

Italians, it was said, could live on less. On the home front, it must been a strange sight to see all the public parks, gardens, and flowerbeds transformed into "war gardens" (*orti di guerra*). They were planted with wheat and vegetables, and local citizens were encouraged to help with the harvest. On the war front, Italian soldiers had to make due with a daily ration that was smaller than that in World War I. At least they had grated cheese and tomato paste with their pasta, along with regular amounts of meat, bread, and wine. But the regulations ominously specified "when available" and stressed the need to source things locally, which was not easy to do at the front. The basic ration soon consisted of canned meat and hardtack.

Even decent tomato paste was becoming hard to find. Owing to the shortage of tin, producers had to return to more traditional methods. Back came the *conserva nera*, manufactured and sold in loaves or sheets.

But this time it was marketed as "sextuple tomato concentrate," outdoing triple concentrate (which really existed) by a factor of two. "Sextuple concentrate" was as hyperbolic and as false as the regime itself.

In the same year, 1942, other foods were rationed: potatoes, legumes, milk, eggs, and cheese. At shops and in markets, Italians queued for hours, never sure if anything would be left when it was their turn. As one Roman housewife remembered: "I would look for onions because with onions you can cook anything, even without oil. But after I queued, maybe for hours, the onions would be finished, and so you joined the queue at another stall and bought what was there, peppers or tomatoes. And that's how the morning would go." Nothing was wasted. For instance, a bit of mustard could be diluted in the liquid leftover from slicing tomatoes for a salad to make a salad dressing.

Italians began to substitute their purchases with recourse to the black market. At first, this meant buying "under the counter" from their usual shopkeepers. But as the war progressed (if that is the right word), urban Italians increasingly bought from peasants who came into town or made trips outside the city limits, to the producers themselves. Rural relations were a godsend. After 1943, all Italians, whether they lived in the Nazi-occupied north of the country or the Allied-liberated south, were faced with high food prices and shortages. "The animal struggle for existence governed everything," according to the journalist Alan Moorhead, writing from Naples. "Food. That was the only thing that mattered." When Rome was liberated the following year, almost three-quarters of the food was available only on the black market.

Advice abounded. The deleterious effects of the war on dietary standards are evident in the cookbooks of this period. Amalia Moretti Foggia, who went by the name "Petronilla," began to offer dietary advice in 1929 in the weekly magazine *La Domenica del Corriere*. In *Ricette di Petronilla per tempi eccezionali* (1941), Petronilla provided suggestions for rationed foods and went shopping with coupons to make the most of her purchases. Yet she still felt able to include items like chickens, rabbits, and garden produce. Two years later, Petronilla's *200 suggerimenti per . . . questi tempi* (1943) went even further with advice on how to economize. In addition to the recipes, she offered suggestions for using old matches, rope, cabbage stalks, apple and orange peelings, chicken guts, and water in which the rice had been cooked. Realistic common sense replaced the previous autarchical spirit. In 1943, the once dependable review *La cucina italiana* published its last issue—somehow appropriately—offering advice for using potato peelings, lupines, acorns, and cauliflower and cabbage stalks.

Finally, Petronilla's *Desinaretti per . . . questi tempi* (1944) provides simple recipes for "light meals" such as soups and the occasional dessert at a time when the black market was a fact of life for everyone. Petronilla was realistic about this necessity, such as buying precious things like eggs, oil, and potatoes directly from farmers. The book begins with the heartfelt wish that "for poor humankind—now bloodied, tortured, exiled, and despairing—will begin the new, long-lasting, unending era of world peace."

7

The Tomato Conquest

Between 1953 and 1956, the journalist and writer Guido Piovene traveled the length and breadth of Italy in a motor car. RAI (Radio Audizioni Italiane), the state-owned public service broadcaster, had commissioned Piovene to describe and report on the state of a country undergoing rapid change. In spirit and approach, the result was not unlike *English Journey* (1934) by novelist and dramatist J. B. Priestly. A native of Vicenza, Piovene had been a foreign correspondent in both France and the United States, so he came to his Italian "grand tour" with a native's knowledge and an outsider's curiosity. He referred to himself as a "traveler-diarist" and described in radio broadcasts his impressions of the people and places he encountered along the way, in a style that was part investigative journalism, part chronicle, and part essay.

Piovene's *Viaggio in Italia* (1957) offers a snapshot of a country on the eve of the great "economic miracle" of the late 1950s and early 1960s, which would transform Italy. Some of Piovene's encounters tell of an unchanging, impoverished, peasant Italy. Visiting a group of hovels in the small town of San Cataldo (Basilicata), Piovene observes the domestic scene:

> A sole pallet serves the whole family; getting undressed is an unknown custom; the chickens' excrement on the few pieces of furniture seems

to be accepted as part of the normal decoration, alongside the peppers and lard, both ornamentation and foodstuff. The peppers, lard, tomato preserve, and, more rarely, pasta; these are eaten sitting in a circle by fishing out of a central dish with tin forks.

On a feast day in the Sila mountains of Calabria, all that the town of Taverna had to offer visitors for food, despite its promising name, were large, flat biscuits and "loaves of bread sprinkled inside with tomato sauce."

But Piovene also provides examples of development and prosperity, and not just in the north. Land reclamation on either bank of the Sele River, in the province of Salerno, had permitted the expansion of intensive agriculture, mostly fruits and vegetables. The "old food industries" benefited the most, Piovene notes, especially tomato processing, for which Salerno was famous. The Naples area boasted numerous industries—steel in Bagnoli, textiles, and, not least, food—"like the Cirio plant, whose products cover the whole range of diet and is the largest in Europe." Farther north, in the province of Parma, Piovene praises the Barilla pasta factory, whose owner also was a technician, "expert in market surveys." Piovene applauds the "American look" of a factory making tomato sauces. The Althea plant, partly owned by Cirio, produced its Sugoro tomato sauce and "instructed tens of thousands of housewives." "It numbers among our fine industries," Piovene enthused, "structured not very differently from those that I have seen in America, with the same hygienic concern, the same careful study in the choice of ingredients, the same laboratory-clean appearance" (figure 34)

The three decades following the end of World War II were years of momentous change in Italian society, bringing rapid and extensive industrialization and urbanization. Regional cultures crisscrossed north and south as a result of internal migrations, all affecting Italians' dietary habits and patterns. Likewise, new methods of food preservation, presentation, packaging, and marketing spread as well. Supermarkets slowly began to compete with small shops, and multinational food corporations with local producers. Although these dietary changes

Figure 34 The four shiny *boules* give this tomato-processing plant the kind of "laboratory-clean appearance" that so impressed the visiting journalist Guido Piovene. The interior is of Baiocchi, Valentini and Company, Vigatto (Parma), early 1950s. (From Pier Luigi Longarini, *Il passato . . . del pomodoro* [Parma: Silvia, 1998])

were both qualitative and quantitative, paradoxically, they appeared to change very little.

The closing years of World War II had brought Italy to its knees, dividing the country between the south, which had been liberated by the Allies, and the north, which still was controlled by the Nazis. Everywhere the economy was in tatters, and hunger was part of daily life. If Italians were not starving or even severely malnourished, it was only because of food relief. This began in 1944, but only in the Allied areas, under the auspices of the United Nations Relief and Rehabilitation Administration. The program began in earnest the following year, and by war's end, it was distributing some 2 million food rations each day to Italian children. In addition, a local and private provisioning network fed millions of hungry Italians. The municipal welfare boards (*enti comunali di assistenza*) ran soup kitchens (*cucine popolari*) and distributed money, food parcels, clothing, and medicines. In 1959, fifteen years after the end of the war, 4 million Italians were still receiving some form of assistance from their local boards.

Even after the emergency had passed, Italy remained poor relative to the industrialized West. For example, its per capita gross domestic product was less than 33 percent that of Switzerland. In addition, in Italy the average salary was 60 percent that of West Germany, 40 percent that of France, and 14 percent that of the United States. An Italian parliamentary inquiry of 1952/1953 reported that on average, Italians consumed 50 percent of the sugar eaten by the French and 33 percent of that enjoyed by the British. Their meat consumption was only 25 percent that of the British and 20 percent that of the Danish. Indeed, 38 percent of Italians—some 4.5 million people—never bought meat at all, and another 27 percent bought it just once a week. Although these figures did not take account of any animals raised domestically, like chickens or rabbits, they were shocking nevertheless. In the south, bread and field greens, along with other vegetables, remained the staples.

The Italian food industry was in a fragile state, tied as it was to the poor condition of the country's agriculture. A report for the Constituent

Assembly in 1947 commented on the decline of the tomato-canning industry "in the last few years." The report linked the drop to the limited availability of fresh tomatoes, as well as the fact that more and more Italians were preserving their own. The immediate postwar years were the most difficult for Italians, as the habits of restraint, austerity, and self-sufficiency proved to be deeply ingrained.

Ninety percent of Italian homes lacked one or more of the modern amenities, such as electricity, running water, or a toilet. Their kitchens usually consisted of a fireplace or a single burner where soup or beans might be cooked. Kitchen ranges, known as *cucine economiche*, had been available since the 1930s, but they still were beyond the means of most Italians. In the early 1950s, the kitchen of middle-class rural families was still basic, furnished with only a breadboard, a flour chest, a wooden table and chairs, a cupboard, a pasta board, and various pans, glasses, and an earthenware pot for cooking legumes. Their diet was equally simple: dishes requiring only basic preparation.

This was the country that the Christian Democratic Party proudly represented as "frugal and peasant"; the Italian Communist Party derided as "poor, depressed, and colonized by U.S. capital"; and the far-right Italian Social Movement nostalgically saw as "destined for misery after losing its 'place in the sun.'" In the tense climate of the immediate postwar years, hunger was an important political issue. The Christian Democrats promoted their ability to put food on Italy's tables, with the generous assistance of the United States. A poster from the hard-fought election campaign of 1948 between the Christian Democrats of Alcide De Gasperi and the Communists of Palmiro Togliatti warned Italians: "Don't think you can season your pasta with Togliatti's speeches. So intelligent people will vote for De Gasperi—who has obtained free from America the flour for your spaghetti and the sauce to go with it."

The shattered and poor country of the immediate postwar years was depicted in the neorealistic films of the time, like *Ladri di biciclette* (*The Bicycle Thief*, 1948) and *Miracolo a Milano* (*Miracle in Milan*, 1951), both directed by Vittorio De Sica. It also turned to its sense of humor—but only after the worst had passed. In *Totò a Parigi* (*Toto in Paris*, directed by Camillo Mastrocinque, 1958) the eponymous protagonist boasts: "I am a genuine starving wretch; mine is an atavistic hunger. I come from a dynasty of starving wretches: my father, my brother, my great-grandfather, my great-great grandfather, my great-great-great grandfather, all the ancestors in my family and collateral branches." In his many films, the tragicomic Totò, developed and played by the Neapolitan actor Antonio De Curtis, became an icon of the hungry Italian, past and present. It was natural, then, for many people to assume that Italy would remain the poor and proletarian country it always had been.

Instead, beginning in the mid-1950s and in less than two decades, Italy ceased to be a predominantly peasant country and became a major industrial nation. The country's potential was finally realized. Several factors lay behind this "economic miracle": vast sums of money were made available by the United States through the Marshall Plan. The end of economic protectionism after Fascism boosted modernization and Italy's capacity to compete. New sources of energy were developed; the steel industry was transformed; and the infrastructure, like expressways, was upgraded. Besides monetary stability, and perhaps most important, the cost of labor remained low.

As an example of the extent of the country's manufacturing boom, let us consider refrigerators. In 1951, Italy produced just 18,500, but six years later the figure was 370,000, and by 1967 Italy was producing 32 million, making Italy the third largest producer of refrigerators, after the United States and Japan. Moreover, between 1958 and 1965, the number of families who owned a refrigerator rose from 13 to 55 percent.

Italians were putting their refrigerators to good use. By 1952, Italians were consuming as much food as they had before the outbreak of World War II, and food consumption kept rising in both quantity and quality, in an uninterrupted series of increases and improvements. For the first time, the majority of Italians had the freedom to choose what they wanted to eat.

They exercised this freedom in interesting ways. The consumption of maize, in the form of polenta, plummeted, as it was too closely associated with centuries of poverty and malnutrition. Between 1965 and 1969, it was only one-third of what it had been between 1951 and 1955, down from 48.4 to 16.9 pounds per person a year. Tomato consumption, by contrast, doubled in the same period, rising from 43.1 to 88 pounds per capita. This was substantially more than the overall increase in fruit and vegetable consumption, which itself went up by more than 50 percent, and meat consumption almost tripled. Food imports increased dramatically to meet this demand, rising to 33 percent of the nation's imports. At the same time, as a proportion of annual income, domestic expenditure on food fell below 50 percent for the first time, to 37 percent between 1965 and 1969.

As many as 3 million Italians, from government employees to factory workers, also regularly ate in company cafeterias, where they got a substantial meal consisting of a pasta or soup course, followed by a second course of cheese, eggs or meat, with vegetables on the side, for next to nothing. According to one union member writing in 1953, at the Marelli factory in Sesto San Giovanni (Milan), workers "eat very well: risotto or pasta with sauce, well seasoned; a varied second course and good wine, too." Moreover, "there are workers whose home is close by and yet still prefer to eat at the canteen." High praise indeed, given the symbolic importance of the midday meal at home. More and more Italians thus became accustomed to eating at least one of their meals outside the home.

Italians put all this food to good use, growing an average of 1.6 inches from 1951 to 1972, to 5.7 feet. But they did not grow fat. Indeed, what is most striking is that even though Italians could now afford to, they did not turn to a high-fat, high-protein diet. Of course, they did buy more meat, dairy products, fats, and sugar, and they ate more refined grains (in white bread) at the expense of rye, barley, and maize. The consumption of legumes dropped. Italians also started to buy their foods in different ways. The first supermarket opened in Milan in 1957, to much consternation. Customers (and the curious) were surprised to see so many different products and goods in the same place, but they did not like the idea of self-service, helping themselves without the intervention of the shopkeeper.

Small, family-run shops and town-center markets remained the norm, and Italians maintained a traditional balance in favor of carbohydrates and vegetable protein—in other words, pasta and bread, vegetables and fruit. Although they were buying a lot more food, the content and structure of their meals did not seem to change much, and they prepared their food in familiar ways.

The American foreign correspondent and food writer Waverley Root commented, and it was meant as a compliment, that "Italian cooking has remained basically amateur cooking even when it is executed by professionals." A dish as simple as "meatballs in tomato sauce" could be elevated to literary status. In a short story by that name ("Le polpette al pomodoro," 1957), the poet Umberto Saba tells his daughter of the love that he and her mother, Lina, felt for each other, a love that she expressed through the cooking of her "favored dishes." "If I tell you that her meatballs were love, I would not be telling you anything new," Saba writes.

In the postwar years, *La cucina italiana* was resuscitated and returned to advocating basic Italian foods. Joining it was Vera Rossi Lodomez and Franca Matricardi's *Il cucchiaio d'argento*, a cookbook that first appeared in 1950 and became a staple of kitchens throughout the country (and a common wedding present). What was new was not so much the content but the notion that women would now be trying to maintain standards on their own, without the help of servants.

Italians remained attached to the cooking of the past, in both structure and content. The paradox was that many people were eating this way for the first time. What we now consider the typical Italian meal structure—a *primo piatto* of pasta, rice, broth, or soup; a *secondo piatto* of fish, meat, or cheese, plus one or more *contorni*, or vegetable side dishes; followed by a dessert of fruit or a sweet—was only now becoming the norm for the majority of the population. In an upwardly mobile society, a family's ability to adhere to this formula was a sign of status and belonging.

Amid all this abundance, though, renunciation and thrift were still regarded as national virtues. In Elio Vittorini's neorealist novel *Conversazione in Sicilia* (*Conversation in Sicily*), the protagonist, Silvestro, returns to his native Sicily on a quest for "lost human kind." Food brings back childhood memories for him, flavors lost and found again. These were simple, peasant foods: herring, peppers during the summer, broad beans with cardoons (edible thistle stalks), and "lentils cooked with onion, dried tomatoes, and bacon . . . and a sprig of rosemary too." His mother recalls that he always wanted a second helping—indeed, he would have given his firstborn son for a second helping—which, needless to say, he never got.

The novel, written in 1936/1937 and first published in 1941, in the depths of Fascism, met with popular and critical success in the 1950s, when it was paired with drawings made for it by the Sicilian artist Renato Guttuso. By then, Silvestro's mother would have been able to let him have a second helping of lentils, and more.

In *Conversazione in Sicilia*, Silvestro's mother reminds him how she dried tomatoes at home for the family. This was just as Italians, especially in certain areas of the south, had been doing for a hundred years or more. But in other ways, the tomato's place in cooking was changing and quite dramatically so. First, it was associated with an unparalleled rise in pasta consumption. In 1947, the Neapolitan writer Giuseppe Marotta wrote of spaghetti as "the ideal food for the person who has toiled from morning to night." Marotta fantasized about becoming a cloistered monk, forever enclosed in the monastery of a spaghetti factory, with its endless blue packets of pasta, bringing with him only cans of tomatoes. From middle class to working class: the dream was at last coming true. Italians could now eat *spaghetti al pomodoro* to their heart's content, and they did, to the extent that stereotype and reality began to fuse. What would Marinetti have made of it?

In the same year, 1947, the Neapolitan playwright Eduardo De Filippo wrote a poem in which the fussy narrator reminds his listeners and his long-suffering wife how to make a proper *ragù*. This was the sauce

based on cheaper cuts of meat and tomato paste and cooked at a low
heat for hours, first detailed by Pellegrino Artusi in the late nineteenth
century and becoming Sunday "gravy" in the United States. If it was not
cooked properly—it had to *peppiare*, or simmer slightly, to use a beautifully onomatopoeic Neapolitan word—it was not *ragù*, but just meat
with tomatoes:

The Meat Sauce

The meat sauce I like
Only Mamma knew how to make.
Since I married you,
We talk about it for the sake of it.
I'm not hard to please;
But I'd rather you didn't make it anymore.

Yes, all right, as you like.
Do we have to fight about it?
What do you think? This is a meat sauce?
And so I'll eat it for the sake of it . . .
But will you let me say something? . . .
This is meat with tomatoes.

['O rraù

'O rraù ca me piace a me
m' 'o ffaceva sulo mammà.
A che m'aggio spusato a te,
ne parlammo pe' ne parlà.
Io nun songo difficultuso;
ma luvàmmel' 'a miezo st'uso.

Sì, va buono: cumme vuò tu.
Mò ce avéssem' appiccecà?
Tu che dice? Chest' 'è rraù?

E io m' 'o mmagno pe' m' 'o mangià . . .
M' 'a faja dicere na parola? . . .
Chesta è carne c' 'a pummarola.]

De Filippo transformed the poem into the focus of one of his most successful plays, *Sabato, domenica e lunedì* (*Saturday, Sunday, and Monday*, 1959), in which the ritual of the Sunday meal, the preparation for it, and the aftermath famously become the centerpiece of a family fight.

In both Britain and the United States, Italian food already was synonymous with spaghetti and tomato sauce. In 1950s Britain, it was still mysterious and exotic enough that in 1957, BBC television could get away with broadcasting a short documentary on that year's bumper "spaghetti harvest." Amid scenes of "spaghetti trees," it referred to the "spaghetti plantations in the Po valley," the fortunate disappearance of the nasty "spaghetti weevil," and the achievements of plant breeders in developing new varieties with equal-length strands, which facilitated harvesting. The date of the broadcast, April 1, ought to have given the game away, but many viewers still were fooled.

Elizabeth David, whose book *Italian Food* had introduced some British cooks to the subject a few years earlier, wanted to do away with the stereotype. She commented that "only the very credulous would suppose that Italians live entirely upon *pasta asciutta* and veal escalopes." Italian restaurateurs and waiters might expect foreigner travelers to want only "spaghetti in tomato sauce, followed by a veal cutlet," but David knew better. She was right to stress the variety of Italian food, in all its regional and local splendor, and her book rightfully remains a classic. The irony was, however, that it was just this kind of meal to which millions of Italians were then aspiring and replicating.

These aspirations are most memorably depicted in the films of the period. This desire for food can be found in *Miseria e nobiltà* (*Poverty and Nobility*, directed by Mario Mattioli, 1954), starring, of course, Totò. In the film, the brothers Felice and Pasquale pawn an overcoat in order to purchase the necessary ingredients for a decent meal: spaghetti with sausage sauce, followed by eggs with fresh mozzarella. Felice (Totò) gives Pasquale detailed and lengthy instructions on the choice of the ingredients, which themselves are part of the anticipation. When the spaghetti is finally served, "full of sauce" according to Felice's instructions, it is the realization of a dream. After their long wait, the famished diners lose their inhibitions, grabbing at the spaghetti with their hands and shoving it into their mouths, with Felice even filling his pockets.

In another film from the same year, another memorable image appears, perhaps the best-known food scene in all Italian cinema (famous enough to be available as a clip on YouTube). In *Un Americano a Roma* (*An American in Rome*, directed by Steno, 1954), Alberto Sordi plays Nando Moriconi, a working-class teenager infatuated with all things American. In one scene, Nando prepares his idea of an American supper—a slice of bread topped with a mixture of jam, mustard, yogurt, and milk—only to spit it out in disgust (no wonder) in favor of a heaping plate of spaghetti with tomato sauce his mother has left for him (figure 35). American culture and optimism might be fine, the film seems to say, but the food has to be Italian.

Sophia Loren, a symbol of vivacious, curvaceous Italian womanhood, once told an interviewer: "Everything you see, I owe to spaghetti." Her mixture of memoir and recipes, *In cucina con amore*, published in Italian in 1971 and in English the following year, is testimony to the continuing earthiness typical of Italian postwar cookery. Writing at a hotel in Geneva, Switzerland, while she was expecting her first child, Loren looked back on her poor childhood (as Sofia Scicolone) in Pozzuoli, on the Bay of Naples. She praises her grandmother for teaching her how to make monotonous and cheap foods taste good and varied, and she provides a recipe for spaghetti with tomato sauce, a version of which is "handed down from mother to daughter in every Neapolitan household."

The change in tomato consumption was first and foremost one of quantity, but just as important was the change in quality. Many Italians could remember a time when the tomato "was hated, as heralding the worst

Figure 3 5 The teenager Nando Moriconi (Alberto Sordi), lover of all things American, feeds his Italian side in *Un Americano a Roma* (1954).

of ills," in the words of writer-painter-composer Alberto Savinio. Even the Tuscan peasants who cultivated tomatoes refused to eat "the 'red' soup because they believed it was harmful to health," according to Ettore Magelli. Many Italians also could remember when preserving tomatoes meant days of toil in the hot sun or kitchen.

Now women could choose to buy processed tomatoes if they so desired, along with the newer, ready-made pasta sauces and condiments, to go with their new kitchens. Old-style *conserva*, formerly sold loose by weight and wrapped in wax paper, was harder to find in shops. Instead, tomato processors launched new products: tomato concentrate in tubes and cubes, and ready-made sauces with mushrooms or meat. Products like these allowed postwar Italian women to become urban, stay-at-home *casalinghe* (housewives), mixing motherhood, domestic cleaning, and meal preparation at a time of increasing expectations. Convenience foods, like "labor-saving" appliances, represented the sort of change that allowed things to remain much the same for three decades. The figure of the homemaker, as both an ideal and a reality, helped absorb the social upheaval of the period.

The tomato was also being used in new ways, appearing in different guises and finding its way into dishes where it never used to be. Just one example among the many possible: the famous Roman dish *bucatini all'amatriciana*. Bucatini, a long tubelike pasta, is served with a sauce of onion, bacon (*guanciale*, from the pig's cheek or jowl, to be precise), black pepper, and tomatoes, with Pecorino cheese grated on top, and is said to have originated in the town of Amatrice, in northeastern Lazio. At one time, it contained no tomatoes, a "white" sauce that survives today (or that has been rediscovered) as *amatriciana bianca*.*

Not everyone thought that creeping "tomatoization" was a good thing. It was nothing less than an invasion. Writing shortly before his early death in 1961, Richard Bethell, fourth Lord Westbury, a long-time resident in Rome and all-around bon vivant, bemoaned the tomato's conquest of Italian cookery in no uncertain terms:

> Had the tomato stopped its advance as a sauce for pasta asciutta, I would have little to say against it. But its conquest has been complete, and it is now particularly impossible in central Italy to eat soup, fish, flesh or salad without the ever-present tomato drowning every other taste.

We cannot imagine Lord Westbury being overly enthusiastic about eggs served with tomatoes.

---

* The effects were just as pronounced in Spain. It is hard to imagine gazpacho—that ingenious chilled "soup" made of tomatoes, bread, oil, vinegar, and garlic—without tomatoes. But, in fact, it started out as a simple way of using stale bread, softened in oil.

Eggs and Tomatoes (*Uova al piatto con pomidoro*)

Remove the skins from 1 lb. of tomatoes. Into a shallow, two handled egg dish pour a small cupful of olive oil, and in this fry a sliced onion. When it is golden add the chopped tomatoes and stew them for about 15 minutes, seasoning with salt, pepper, garlic if you like, nutmeg, fresh basil or parsley. When the tomatoes are reduced more or less to a pulp break in the eggs and cover the pan. The eggs will take about 6 or 7 minutes to cook and should be left until you see that the whites are set and the yolks still soft. From the moment the eggs are put in, the dish can alternatively be put, covered, in a medium hot oven.

Elizabeth David, *Italian Food* (1954; repr., Harmondsworth: Penguin, 1963), 143.

Elizabeth David might have been "charmed" by the vivid color that tomatoes provided. But she, too, included the "too frequent appearance of tomato sauce" as one of the "faults of Italian cooking" (along with the excessive use of cheese and powerful herbs like rosemary). Other food writers agreed. Waverley Root noted in 1971 that the tomato "is ubiquitous in Italian cooking today." Root probably would have preferred the Italian cooking of "yesterday." In his section on the food of Florence, for instance, tomatoes appear in a wide variety of dishes from tripe to minestrone and from mutton stew to liver. But aside from an obligatory reference to the production of tomato paste in the Parma region, the tomato industry is conspicuous by its absence from Root's book, as if he were trying to exorcize its presence in his search for "pure" regional food.

Tomatoes do make a brief appearance in Root's chapter on Sardinia, drying on rooftops, to be used instead of paste, "a manufactured product too expensive for poor Sardinia." In the chapter Root dedicates to the Campania region, tomatoes appear in the form of *melanzane alla parmigiana* (eggplant Parmesan), where the "Parma" could perhaps refer as much to the tomato paste as to the Parmesan cheese. In fact, Parma, or any of its food products, may have had little to do with the origins of the dish as prepared in the Italian south. Some purists today prefer to call the dish *parmigiana di melanzane*, arguing that the *parmigiana* actually refers to its layers, derived from the Sicilian word for the slat of a shutter (*parmiciana*).

The misgivings of gastronomes like Bethell, David, and Root aside, and regardless of the raging debates over nomenclature, tomatoes were spreading, for better or worse. All the tomatoes, fresh and processed, that Italians had been buying since the 1950s had to *go* somewhere. So where were they *coming* from?

On the surface, the tomato industry in Italy seemed to remain quite traditional. Tomatoes still were harvested predominantly by hand. Canning factories still tended to be located close to the fields, even if the tomatoes were now "sent to their destiny aboard giant trucks," in the words of the Neapolitan writer Domenico Rea, rather than in "those heavy carts, drawn by horses."

But in the postwar decades, there were enormous changes here too, in both quantity and quality. Most of the quantitative increase came in the late 1950s and early 1960s, before tailing off slightly during the 1970s. Italian tomato production rose by 75 percent from 1957 to 1967, much more than any other crop. In 1967, the country harvested some 3.9 million tons. Land dedicated to tomato cultivation increased from 271,816 acres in 1957 to 321,237 acres ten years later. Productivity increased, too, with the quantity of tomatoes produced per rising steeply from 1957 to 1967.

These national statistics hide important shifts at the regional level. In the region of Emilia Romagna, tomato cultivation actually declined during this period, and production fell, particularly during the 1970s, by some 20 percent, as did the amount of land devoted to tomato cultivation. The increase in tomato cultivation and production was thus mostly in the south, with production in the regions of Campania and Puglia nearly doubling.

Puglia also was the chief beneficiary of rising rates of productivity. In 1959, an acre in Emilia Romagna produced almost three times as many tomatoes, by weight, as did an acre in Puglia. This disparity narrowed during the 1960s but then rose in the late 1970s, so that by 1984 the three regions had similar productivity rates. More remarkable still, an acre of land in Puglia now produced on average almost three times as much as it had in the 1950s. Nevertheless, Puglia was still sending most of its tomato crop to Campania for processing, much as it had done since the 1880s. The bulk of the south's canning plants were in Campania, where, in the late 1960s, 164 plants employed some 1,500 people during the canning season, even though most of the factories remained relatively small-scale operations.

Italians consumed large amounts of tomatoes fresh, or freshly cooked, but most of the tomato crop was processed in some way, primarily into peeled whole tomatoes or tomato concentrate. The change instead was who was buying them. Before World War II, increased production would have been reflected in increased exports abroad. Now, as we have seen, Italians were consuming tomatoes in greater quantities than ever before. Exports of processed tomatoes did rise between 1957 and 1967, but "only" by 25 percent. But tomato production rose by 75 percent during the same period, feeding the Italian boom.

The main importers of Italian processed tomatoes remained Britain, the United States, and West Germany. But there was a perceptible shift. Britain and the United States were importing less: Britain presumably because other Mediterranean countries were now competing to supply canned tomatoes, and the United States because of its expanding domestic production, notably in California, which led the world in tomato cultivation and production. The country that took up the slack thus was West Germany. It was no coincidence that by this time, West Germany was also the European country with the highest number of Italian immigrants. For this reason, the increasing German importation of Italian processed tomatoes was, at least in part, an extension of Italian "domestic" consumption.

Tomato productivity increased because of the introduction of new methods of cultivation, including the greater use of fertilizers and pesticides. But, above all, the increase was due to the adoption of new varieties of tomato.

This qualitative shift was a surprise, for it overturns our ideas about "traditional" foods. As early as 1954, Ettore Magelli reported on the virtual disappearance of the 'San Marzano' tomato from cultivation in the province of Salerno. Only the occasional plant could still be found growing amid other varieties, most notably the related 'Lampadina' (Lightbulb), with its smoother skin, more easily removed for canning. Another guide to tomato cultivation, written by Arturo Giordano and published in 1961, attributed this decline to poor technical skills in the region, primarily the failure to select and separate the best and most appropriate seeds. The result was accidental crossbreeding, or hybridism, and the victim was the 'San Marzano' and overall tomato quality.

Whether intentional or accidental, the introduction of different varieties was actually nothing new, as the development, adoption, and processing of new varieties had been part of tomato agriculture and industry since at least the 1880s. Perhaps it was the Fascist period, when there was little change, that was the exception.

In any case, the 'San Marzano' remained a popular variety in Italy, despite what the manuals had to say about it. In 1965, it still represented a bit more than one-third (35%) of the country's tomato production, but by this time it had already been overtaken by a similar, but newer, variety, the 'Roma'. Its name may have been Italian, but it was developed at the Plant Industry Station in Beltsville, Maryland, in the mid-1950s. With the canning industry in mind, plant breeders there crossed the 'San Marzano', 'Pan American', and 'Red Top' varieties to produce a plant that was more productive, more resistant to disease, and had larger fruits than the 'San Marzano'.

Moreover, the 'Roma' is determinate, rather than indeterminate like the 'San Marzano'. This means that the plant stops growing when the fruit sets on the top bud, and the fruits then ripens around the same time, making harvesting easier and reducing labor costs. The 'Roma' also does not require pruning and needs very little staking, for it is a "bush" tomato. Indeterminate varieties like the 'San Marzano', by contrast, grow like vines unless regularly pruned, and they flower, set, and ripen fruit throughout the growing season. It is easy to see why Italian growers, particularly in the south, began to favor the 'Roma' over the 'San Marzano'. By 1969, the 'Roma' accounted for half the Italian tomato harvest; the 'San Marzano', only one-fifth.

Curiously, the 'Roma' became, and remains, one of the best-known varieties in North America and Australia. Its name, shape, and taste helped it become a symbol of "Italianness" in the New World. So successful was it that in the United States. it is sometimes assumed to be a traditional Italian tomato. It is traded as an "heirloom" variety by seed collectors and lovingly grown and bottled by the descendants of Italians in a culinary reinforcement of ethnicity.

In Italy in 1969, for the first time, new pear-shaped, early-maturing varieties found a place—albeit small, only 5 percent—in tomato production. Varieties like the 'Heinz 1706', 'Ventura', and 'Chico' were attractive to growers because they mature over a shorter period. The 'Chico', for instance, is closely related to the 'Roma' but is slightly larger and lower in acidity. It was developed for commercial purposes by the Petoseed Company in Chico, California. These varieties represented 25 percent of the Italian tomato harvest in 1973 and 30 percent in 1977, on the way to supplanting both the 'Roma', down to 30 percent in 1973 and 20 percent in 1977, and the 'San Marzano', 10 percent in 1973 and only 5 percent in 1977.

Another novelty tomato in 1973 was a "square" variety, like the 'Petomech' and the 'Cal-j', resistant to bulk handling and cracking. Two years after their introduction in Italy, square varieties constituted 10 percent of tomato production, rising to 30 percent by 1977.

Could we expect to find some stability in the large, furrowed, round varieties favored in Emilia Romagna? After all, in 1969 they still made up 25 percent of the tomato crop. But there was change here, too. In the

early 1950s, in an increasingly competitive industry, Parma-based producers started to look for less acidic, sweeter varieties. To find a tomato
that was less tart, they turned to the American cultivars that were then
being developed for the "fresh," or salad, market. The preferred variety
was the 'Geneva 11', developed in New York State by Cornell University's
plant breeders. Other varieties developed by Heinz's and Campbell's plant
breeders were grown as well. Even so, round tomatoes accounted for just
15 percent of tomato production in 1977.

If the history of the "Italian" tomato began with its crossing from
the New World to the Old in the sixteenth century and then its return
to the Americas in the late nineteenth century in immigrants' baggage, it
now crossed the Atlantic yet again, in the form of cultivars developed by
American researchers and agribusinesses.

Other factors—and a good deal of chance—were responsible for taking
the tomato into parts of Italy where its place in the diet earlier had been
limited or nonexistent. One factor was the bathtub.

In 2000, a family doctor in Turin told the historian Laura Fiorini that
in the late 1960s, southern Italian "immigrants" in the city would fill their
bathtubs with soil and grow tomatoes in them. It is hard to say whether
the implication that tomatoes were more important to these people than
hygiene is a simple racist slur, an urban legend, or the reflection of a different reality. (From a practical point of view, light levels would be a bit
low in the average apartment bathroom for successful tomato cultivation,
but at least watering and drainage would not be a problem.)

The doctor's memory echoes one of the most significant and trau-
matic social occurrences of the postwar decades in Italy: internal migration. The people to whom the doctor was referring were recent arrivals
from the south, living in cheaply built, high-density, subsidized housing
and employed at the nearby Fiat car factory. But years before the memory
was shared with the historian, this story already was legendary, an infamous indication of the ongoing divide between north and south. The
southern singer-songwriter Mimmo Cavallo, a native of Lizzano (Puglia),

even made growing tomatoes in the bathtub a badge of southern Italian identity in his song "Siamo meridionali" (We're Southerners, 1980).

The prosperity in Italy referred to earlier in this chapter had a pronounced regional dimension. The main beneficiaries of the economic miracle were the northeast—the industrial triangle formed by Milan, Genoa, and Turin—and some parts of the center, the areas where the capital, resources, and expertise were concentrated. The result was a growing imbalance between north and south. During the 1960s, northern Italians continued to spend more on food than southerners did. But because northerners earned more, the sum spent on food represented a smaller portion of their annual incomes. Northerners still consumed more meat, dairy products, fruit, sugar, and fats than did southerners. For their part, southerners consumed more carbohydrates, fish, and dried legumes.

Poor Italians coped as they often had done: they migrated, bringing their dietary habits (and aspirations) and their gardening skills with them. Millions went overseas, for instance, to Canada and Australia. Just as the previous wave of immigrants had been attracted to food manufacture and marketing, 13 percent of Toronto's food retailers were Italian by 1961, and Italians owned 250 food stores and markets in Montreal by the late 1960s. In Australia, Italians and their descendants still grow a significant proportion of the tomatoes. Landownership there was initially made possible by the "share-farming" arrangement, in which the landholder provided the land, irrigation, and housing, and the grower kept two-thirds of the produce.

Other Italians took jobs in other parts of Europe. By 1965, 2 million Italians, three-quarters of whom were from the south, were working in West Germany, Switzerland, France, and Belgium. The Italian government negotiated bilateral agreements with other European countries, linking surplus labor to trade and import agreements. For instance, in 1955, Italy signed a precedent-setting agreement with West Germany that created the guest-worker (*Gastarbeiter*) migrations of the 1960s. We have seen that tomato imports to Germany increased, partly as a result of this immigration. In Belgium, Italians worked in the coal mines, often in terrible conditions. By the 1970s, however, the derogatory label *sales maca-ronis* (dirty macaronis), as Italians there were sometimes called, had been dropped. Both the immigrants and their cuisine came to be accepted, even valued, in the host country. Testimony to this is the appearance of a recipe for Italian pasta *sugo* in the 1972 edition of the much-loved Flemish-Belgian cookbook *Ons kookboek*. It is just one example of what has been called the "Italianization" of Flemish cuisine.

The main difference in the postwar migratory wave compared with previous ones was that most of the movement took place *within* Italy. Out of a population of 50 million, as many as 45 million Italians changed their town of residence between 1955 and 1981. Fifteen million were long-distance migrants, including the 8 million or 9 million southern Italians who moved north. Most left the rural south for jobs in the cities of northern and central Italy, as rural workers could expect to at least double their incomes in factory jobs. They also were attracted by the regular hours and wages. Although peasants had always had more than enough to do during harvesttime, they had few means of earning money during the winter months.

The populations of both Milan and Rome grew markedly during this period, but it was the city of Turin, given its smaller size, that felt the impact of internal migration most strongly. Between 1951 and 1967, its population increased from 719,300 to 1,124,714. Turin also received a much higher proportion of southern immigrants, mainly from the provinces of Foggia and Bari (in the Puglia region) and Reggio Calabria. This influx, of more than 330,000 southern Italians, made Turin the third largest "southern" city in Italy, after Naples and Palermo.

The Turinese were not ready for the migrants' poverty and "strange" habits. Each group regarded the other as foreign and threatening. By the 1960s, one-third of Fiat's employees in Turin were southern born. Workers were tagged by their origins: a Piedmontese was a *minestra* (after their penchant for soup); a native of the Veneto was a *polenta*; and a southerner was a *maccheroni*. Food preferences were an obvious marker, but they also paved the way to the eventual cultural mixing. The company cafeteria was often the place where the assimilation began.

Turin's infrastructure—its housing, hospitals, and schools—was not ready either. Infant mortality rose sharply owing to the unhygienic and overcrowded conditions in which the early migrants and their families lived. The daily effort to establish oneself took priority over comfort. In 1962, one migrant to Turin described his day as "work, spaghetti, and sleep." The realization of pasta as a staple is seen not as a sign of abundance but as mere survival. The poor and crowded living conditions became a scandal. Turin's largest employer, Fiat, argued that housing was not its responsibility, but the city's. Tower blocks were belatedly built, allowing families to own an apartment of their own—complete with bathtub. These developments, like those at Mirafiori Sud, where almost three-quarters of the residents came from the south, were not much to speak of, but they were usually a great improvement over their previous conditions.

To move to the factories of the north, Italians in the south sold their houses and land and abandoned their parents but brought with them packets of pasta, bottled tomato sauce, and dried figs. When the fictitious southern migrants Antonio Capone and his brother Peppino arrive in the capital of the north, Milan, they are dressed in winter parkas, as if equipped for an expedition to Siberia. Once in the apartment where they are staying, Antonio removes from his suitcase a loaf of bread, a whole cheese, and, with great reverence, two packets of spaghetti. On the wall behind them, a prosciutto, a braid of garlic, and a string of chilies hang from nails. The film is *Totò, Peppino e la malafemmina* (*Toto, Peppino, and the Hussy*, directed by Camillo Mastrocinque, 1956), sending up the southern migration that was just beginning.

What the new residents could not bring with them they "imported" from the south. On Sundays, Turin's market at Porta Palazzo was transformed into a meeting place for southern immigrants, especially when truckloads of olive oil or tomatoes arrived. (Today, this practice has been continued by more recent immigrants to the city, for example, from sub-Saharan Africa.) A southern Italian stallholder at Porta Palazzo remembered that the smaller tomatoes common in the south "at first ... didn't exist here [in Turin]. Then we brought them in, and so the southerners knew where to find them." Just as Italians in the Americas had done two generations earlier, southerners fanned out from Turin's wholesale vegetable market each morning, peddling the fruit and vegetables at competitive prices to restaurants, bars, and cafeterias. Other southerners set up their own "importing" businesses, bringing in canned whole tomatoes, apparently "not available in Turinese shops" and apparently not seeing the irony in having to import canned tomatoes to the city of Francesco Cirio's birth.

Gradually, as conditions improved, dietary habits changed, affecting the host city as much as it did the more recent arrivals. Turin, like much of northern Italy, eventually became just a bit "southern" owing to the increased consumption of pasta, pizza, olive oil, and fruits and vegetables.

The southerners in the north changed too, variously following the patterns of continuity, assimilation, or rejection that we encountered in chapter 5. The realization of the *pasta al pomodoro* dream came at a cost. Turinese family doctors were struck by how much pasta their southern patients could eat. That the elderly, especially, ate the same foods as they had in the south, but in greater quantities—in addition to more meats and fats—eventually led to high rates of metabolic diseases, hypertension, diabetes, and obesity.

The southerners' first reactions to Turinese cuisine were decidedly negative. In the words of one southern migrant: "Southerners found themselves eating with Piedmontese, these people who eat broth 365 days a year, and they put a little wine in it, turning it purple, which for someone from Puglia is a bit revolting." Another complained that Piedmontese "foods are more doctored, whereas in the south the foods are poorer but more genuine." This was symbolized by *agnolotti*, the stuffed egg-pasta parcels that are one of Piedmont's great contributions to Italian cuisine but about which one migrant claimed: "I never eat them, on principle, because I don't know what's inside; I'm suspicious."

Not all southerners were so hostile to Piedmontese food. In September 1972, the Fiat company magazine, *Illustrato Fiat*, reported the curious case of one of its workers and his family from Gallipoli (Puglia). The wife admitted that "we've become used to a different taste in food. Southern Italian salamis, for example, are too heavy. The cheese is too sharp. Back down in Gallipoli, for the whole fifteen days of our holidays, I had to cook different meals for my family from the ones my in-laws ate." Not a statistically valid sample, of course, but an indication of the possible different reactions to novelty and difference.

More typical perhaps was the gradual, almost imperceptible, change in ingrained habits. The change was not just geographical, but social and economic as well. Foodways developed in the rural south seemed less appropriate in the modernized, industrialized, urbanized north. As one southern woman in Turin recalled years later, in the late 1990s,

> I have to say that we kept on making preserves. . . . We did them less with the advent of the refrigerator, obviously. Still, even now people continue to make preserves, of tomatoes, for example. As long as my mother-in-law and my mother were able to do it, when we went down south for the summer holidays, the whole family would be busy for days making sun-dried tomatoes and bottles of sauce. Now we don't make them anymore, but often we're given them as presents or we buy them from friends who make them.

Why bother, when Cirio preserves were a household name? Viewers of Italian television were treated to a fifty-second spot for Cirio *pelati* during the popular program *Carosello*. With picturesque views of the Bay of Naples and mandolin music in the background, the voice-over reminded viewers that a can of Cirio tomatoes would bring "the taste of the sun" to their tables. Indeed, the word "sun" was uttered four times during the short commercial.

Advertising prospered from the boom and in turn fed consumption. By the 1970s, the economic miracle had more than doubled Italian incomes, and by the end of the decade, the rift between north and south, at least in regard to food consumption, had all but disappeared. Consumers, apprehensive at first, were inundated with advertisements for Cirio preserves, Buitoni pasta, Knorr soup cubes, and Pavesini biscuits. Most advertising was for food, with "mother" in the starring role (or target), showing the products as part of the modern Italian lifestyle, as important as owning a refrigerator or washing machine.

The television program *Carosello* reinforced the attraction to brand names. It is no coincidence that the program began in the same year, 1957, that Milan's first supermarket opened and that it ended some twenty years later, during the rapid expansion of private television channels and massive advertising. *Carosello* was the advertising window that brought daytime programming to a close and led into the evening programs. For the RAI (now Radio Televisione Italiana), the state broadcaster and still suspicious of advertising, the program offered a compromise. It mixed short films, sketches, and cartoons with commercials, which starred well-known personalities, like the singer Mina praising the virtues of Barilla pasta. By 1960, *Carosello* was the most frequently watched program in Italy, affecting the tastes, aspirations, and choices of millions of Italians.

With hunger banished, the tomato began to assume its place in the much-touted "Mediterranean diet." Fresh tomatoes, no longer eaten out of necessity by Italian peasants and the urban poor as a matter of subsistence during the dog days of summer when little else grew, could now be appreciated for bringing "summer" into the lives of urban consumers. Liguria's mixed salad known as *condijon* (or *condiglione* in Italian) could take its "rightful" culinary place alongside France's "salade niçoise," which it resembles. In the south, varieties of tomato developed for the canning industry or to be hung during wintertime and eaten because they were cheap, were now gastronomic fare, like the 'Tramonti' tomato.

We end on a high note. At the end of the 1970s, Italian tomato cultivation and production began to climb again, following a few years of

---

September Salad

The tomato is really mature. If it isn't harvested, it will fall from the bush. . . . It is in this month that the celebration of the tomato reaches its apex.

The little Tramonti tomato is squat and bellied, hard as a rock, with a gut full and blind. It is ideal for preparing the classic tomato salad to eat together with moistened maize [corn] bread. You slice the tomatoes, sprinkle on some salt, sliced onion, a sea of oil; and by itself, it can keep a person on his feet full of energy till evening. But many people further flavor this classic salad with green chilies, sliced eggplant, and red peppers, with cubes of cucumber and a shower of basil leaves. This food contains all of summer.

Domenico Rea, "I mesi" (1988), in *Domenico Rea: Opere*, ed. Francesco Durante (Milan: Mondadori, 2005), 1483–84.

---

decline, due in part to the oil crisis and resulting recession. Soon there were more Italian tomatoes than ever before, at any time in this fruit's history. Land dedicated to tomato cultivation reached a peak of 358,000 acres in 1984 and, combined with higher-yielding varieties, saw tomato production soar to three times what it had been in 1957.

The reason was subsidies. In 1984, food and agricultural subsidies formed 70 percent of the European Common Market's total budget. Beginning in 1978, as part of Europe's Common Agricultural Policy, all of Italy's tomato-growing regions took advantage of the subsidies negotiated by the Italian government. The subsidies were paid to processors, who passed part of them on to growers. The policy also included tariffs on imported goods and export subsidies to encourage exports. The aim was to guarantee a minimum price for a whole range of agricultural goods. Moreover, subsidies were regarded as a social tool to assist economically deprived areas and preserve the rural environment. In the case of tomato production, subsidies were intended to protect European producers from world competition, primarily that of California, at a time when market prices were very low. In Italy, the main beneficiary was Puglia, still one of Italy's poorest regions. Its production of tomatoes consequently increased by a factor of ten to become the country's highest producing region.

At the time, this growth in food production was seen as a good thing. Farmers and producers earned more; exports rose dramatically; and European consumers paid lower prices.

The Italian tomato had never had it so good.

Epilogue

Is the tomato in Italy a victim of its own success? Pietro Citati seems
to think so. In an article a few years ago in Italy's leading newspaper, *La
Repubblica*, the literary critic, writer, and newspaper columnist waxes nostalgically about the tasty tomatoes of the summers of his youth in Liguria. (Citati was born in 1930.) He remembers that the tomato "constituted the heart of my world." "Not tomato sauce, or rice and tomatoes,
which are already corruptions, but the pure tomato, seasoned with oil
and salt." Citati's tomato was the "supreme fruit of the Mediterranean:
gilded, caressed, loved by the sun, that gave form to its inner meaty pulp,
where my teeth bit into it, its delicate skin, its seeds, exquisite aroma, and
color, worthy of a Chardin or a Veronese." Artists like those may come
and go, but the tomato, Citati suggests, is in irreversible decline.

Nowadays, tomatoes look the same everywhere in Italy. Whereas "the
real tomato has different, complicated shapes, with splits and streaks,
and often pronounced baroque features, which so pleased the Neapolitan
painters of the seventeenth century," tomatoes today taste of nothing;
they are full of water. We have lost much, much more than we realize,
Citati warns: "Previously, the pulp, juice, and color passed into the brain,
bathing it, just as the tomato had been penetrated and bathed by the
sun." He hopes that somewhere in the Mediterranean, perhaps in Africa,
people still are growing the tomatoes of yesteryear. Or maybe some

daring young grower in Liguria, Puglia, or Sicily will come forth and offer them for sale. It would not take much capital, only "excellent seeds, a little water, sun, diligence, care, attention, and a supermarket offer." "Real tomatoes" would be in great demand. Citati himself would be prepared to spend $13 a pound (€20 a kilogram) for them.

Citati's article appeared in August 2006 in the midst of the tomato season and at a time when most Italians were taking things easy. It was meant to stir up controversy, and it did. To my knowledge, no one wrote in to object that Neapolitan painters were not yet painting tomatoes in the seventeenth century. But in letters to newspapers and agricultural magazines and in blogs, people commented on every other point Citati raised. Many agreed with him, mentioning their preferred varieties that had been lost to "progress." Yet there was more to this article than the nostalgia (and the deteriorating taste buds) associated with the aging process. If tomatoes do not taste like they once did, it was because they themselves were different. New hybrids were developed to increase yields, withstand pests and disease, and make harvesting easier, but rarely to taste better. Some blamed the subsidies from the European Union (EU); others, market demands. Growers admitted that they were only following the expectations of the market and the food industry. Other respondents, growers and consumers alike, maintained that was not true. You could still find choice—and "heirloom" varieties, if you were prepared to hunt them out—and pay more. And there's the rub. If people were truly prepared to pay four or five times more for their fruits and vegetables, there would be no problem. But they were not prepared to do so. So why should the supermarkets or the large distributors be interested?

Carlo Petrini, the founder of the Slow Food movement, went even further. In an interview for *La Repubblica*, he stressed the environmental effects of what he called the "overproduction of food." In response to Citati's offer to spend more on "real tomatoes," Petrini noted that Italians were spending a smaller proportion of their incomes on food than ever before, half the figure of 1970, and they could spend even less, say 10 percent, by buying tomatoes from China. But saving money had hidden costs: the Chinese tomatoes were "obtained by slave labor," had "traveled by ship, consuming energy" and polluting, and might result in the disappearance of traditional Italian varieties of tomato. Petrini concluded by calling the overproduction of food the main source of the world's pollution. "We produce food for 13 billion people," he pointed out, "and there are 6.3 billion of us, and 800 million go hungry."

Today's tomato is as complex as the world in which we live, and it is harder than ever to take sides. Who is right, and who is wrong?

Tomato processing is both a specialized and a lucrative business. Using cutting-edge technology, tomatoes intended for tomato paste are cultivated by direct seeding, sown by precision machines using coated seeds, and harvested entirely by machine. The varieties are almost exclusively modern hybrids, with names like 'Perfect Peel', 'Isola', and 'Snob'. (Who was the person at Asgrow who thought up that last one?) Tomatoes intended for canning whole are cultivated slightly differently. Instead of direct seeding, "plug" seedlings are used, and the tomatoes are harvested still predominantly by hand. Only hybrids are grown, such as 'Hypeel', 'Italpeel', and 'Calroma'.

Today's tomato never stands still. Multinational seed companies are hard at work on developing ever newer hybrids, able to withstand drought or give us more of what is "good for us," such as antioxidants like lycopene, supposed to help prevent cancer and heart disease. Although genetically modified tomatoes seemed a real possibility in the 1990s, they turned out to be a dud.

In Europe, tomato concentrate made with genetically modified tomatoes was first marketed in Britain in 1996. But a backlash in the United States against another transgenic tomato, the 'FlavrSavr'—which contained a deactivated gene responsible for producing the enzyme that enabled the fruit to soften—took down the tomato concentrate with it. Even though the American public has come to accept genetic modification in commodities like maize and soybeans, tomatoes were seen as a step too far. Italians' opposition to transgenic tomatoes came from both consumer groups and growers' associations. One frequently stated objection was that farmers would have to buy their seeds each year from multinationals like Monsanto, since the seeds contained in the fruits they grew either would be sterile or would not grow the same tomatoes. (The same has been true since the 1950s for most of the hybrids grown in Italy and everywhere else.) Today, no genetically modified tomatoes are sold anywhere in the EU, either fresh or processed. But Europe's strong biotech lobby has not given up. Research continues on genetically modified tomatoes, ready for the day when consumers are "willing" to accept them. In addition, tomatoes in pill form, enabling us to benefit from their cholesterol-reducing lycopene (especially in tomato concentrate) without actually having to eat them, is already in the testing stage.

Tomato processing is one of Italy's great successes, leading to the export of products proudly bought and sold as "made in Italy" for more than a century. In addition to entrepreneurial know-how, there has been help along the way. The agricultural subsidies introduced by the European Common Market in the 1970s continue, albeit in a slightly different form (although as I write, the entire Common Agricultural Policy is under review). As it stands, under a quota system, Italy's share of the European Union's subsidies for tomato cultivation is slightly more than half the European total. Starting in 2001, Italian growers have received a subsidy from the EU for cultivating tomatoes destined for processing. Recently, however, the subsidy has been half the grower's price, or, to put it another way, once the costs are accounted for (fertilizer and labor), any profit that the growers make for their labors is the subsidy itself.

Subsidies have had the effect of encouraging cultivation for processing. Not all tomatoes end up in tubes, cans, or jars, which explains why so many are sold and consumed fresh locally. In addition to payments to growers, the EU also provides millions of euros in export subsidies, which enables Italian producers to export at low prices. Not surprisingly, California producers—and the state is the world's largest tomato processor—would like to see an end to these subsidies. Although the United States government pays the country's maize farmers an astonishing $25 billion a year (representing around 80 percent of their earnings) and more than $1 billion to rice growers (representing the total value of their crop), the growers of fresh produce, including tomatoes, receive no direct support. The countries hardest hit by the EU subsidies, however, are those of West Africa because Italian producers have been able to flood the market there with cheap tomato paste, undercutting, for example, Ghanaian producers.

In addition to the subsidy, the EU enforces a tariff of 14.4 percent on imported tomato products, to make them more expensive, and the World Trade Organization prohibits developing countries from responding in kind if they want its assistance. So Ghana cannot protect its fledgling industry, even though developed countries can. For instance, in order to make its own processed tomatoes competitive with subsidized Italian tomatoes, the Australian government also imposes a tariff on them, although at 5 percent, it is relatively low.

Tomato products thus are very big business in Italy. Although the country exports far more than it can produce from locally grown tomatoes, it makes up the shortfall by importing processed (or partly processed) tomatoes from other countries, such as Turkey and China. Italian growers, and much of the press too, have been extremely hostile to Chinese imports, often using expressions like "yellow peril." In 2004, Italy imported 165,000 tons of tomato concentrate from China, worth $88 million. Because the tomato concentrate was repackaged in Italy, the label could read "made in Italy," and because it was imported in unfinished condition, it was exempt from import duties. This "country of origin" legislation passed in 2006 was meant to outlaw this sort of ambiguous labeling, but in 2004, tomatoes rotted in Italian fields because it was not commercially viable to harvest them.

The irony in this is that the Chinese, apart from their increasing consumption of fast foods, seldom eat tomatoes. The origins of "ketchup" may have been Asian, but the condiment never contained tomatoes. The Mandarin word for "tomato" is 番茄, transliterated as *fanqie* (literally, "barbarian eggplant"). Finally, in the 1990s, Italian producers, eager for a cheap source of processed tomatoes, helped construct factories in China and advised on cultivation. The Chinese were happy to oblige and now export processed tomatoes on their own to Europe and Africa, and are now the world's largest producer of tomato concentrate.

These subsidies and cost cutting were not enough to save Italy's Cirio Company from bankruptcy in 2003, taking down thousands of investors with it. Until then, Cirio had managed to retain a place in Italian hearts by positioning itself as a traditional company, tracing its roots back to its founder, Francesco Cirio, in the mid-nineteenth century. In reality, though, it had long ago ceased to be a family company, and in any case, processed tomatoes were a later addition to the company's repertoire. Cirio was declared insolvent after defaulting on more than $1.4 billion in bonds and was dismantled and sold. In 2004, the company's Italian branch was purchased by a consortium, led by the rival food-processing company Gruppo Conserve Italia, based in Bologna. The Cirio brand was relaunched in 2008 in a television commercial featuring the famous French actor Gérard Dépardieu using Cirio canned tomatoes to prepare a pasta sauce for his expectant guests. Dépardieu beams proudly at the camera, boasting in hybrid Franco-Italo-Neapolitan, "Tengo 'o core italiano" (I have an Italian heart). The commercial's message was at once international and local, modern and traditional, with a hint of passion and southern Italian sunshine.

Italy has never produced a film to rival *The Attack of the Killer Tomatoes!* (1978), directed by the Italian American director John De Bello. In this parody of Hollywood B-movies, a housewife peeling some tomatoes triggers the rebellion of a multitude of large, angry tomatoes intent on conquering the planet. For Italians, the reality of work in tomato fields probably is horrific enough. But worse was the organization and exploitation of Italian labor by local gang bosses, depicted in Michele Placido's film *Pummarò* (1989). The film tells the story of a Ghanaian medical student, Kwaku, who travels to the Caserta area (Campania) to find his brother, a tomato picker nicknamed Pummarò, who has disappeared.

The harsh conditions of the tomato pickers, many of whom were African immigrants, are vividly depicted. In scenes evocative of depictions of African American slavery, fields of cotton are replaced by fields of tomatoes. *Pummarò* was one of the first Italian films to recognize that Italy—for so long a nation of emigration and, more recently, of internal migration—was now on the receiving end, with immigrants performing the tasks that Italians now shunned.

Since the expansion of the EU in 2004, eastern Europeans have taken the place of Africans. In the summer of 2006, when Citati launched his tirade against flavorless tomatoes, an investigative journalist went undercover to explore the conditions of immigrant tomato pickers in the vast, treeless plain south of Foggia (Puglia), known as the Tavoliere. Fabrizio Gatti had investigated the terrible humanitarian conditions of the Italian refugee camp on the island of Lampedusa, passing himself off as a refugee and experiencing life there firsthand. The following year, once again for the respected Italian newsmagazine *L'espresso*, Gatti disguised himself as an immigrant tomato picker for an investigation that won him the European Union's "journalist of the year" award.

Pretending that he was a Romanian was easy. For between $21 and $28 a day, Gatti worked under a ruthless gang boss. Lodging, for which Gatti had to pay $7 a night, was a disgusting makeshift shack without water or electricity. If the workers showed up late for work in the fields, they would be beaten. If they missed a day, even if they were sick, they had to pay a fine of one day's labor. The farm owners and tomato buyers either ignored such abuses or anonymously tipped off the local police to the presence of illegal immigrants working in the fields, typically on payday, so that the police raid relieved the boss from having to pay them. Even Doctors Without Borders set up a mobile hospital in the area, like those used in war zones in the developing world. Workers also "disappeared." Gatti finally had to flee on a bicycle when he was told that his gang boss was angry with him.

In its underlying violence and inhumanity, the Tavoliere's labor system has changed little from the days of the landless day laborers, locally called *giornatari*, who assembled before dawn each day in the central square in the hope of securing a day's work. A hundred years ago, the main crop in the Tavoliere was wheat. The overseer, or gang boss, known by the military-sounding title of *caporale* (corporal), hired the workers on behalf of the estate owner. He rode on horseback and carried a long thick stick, pointing at the *giornatari* he wished to hire and leading them to the distant fields and sometimes beating them with it if they were slow. The *giornatari* of the Tavoliere were in an even worse position then Parma's day laborers (examined in chapter 6) because they had no system of tenant farming to go back to. All they had was year after year of hired toil in different places, for no set wage and no job security. When the *giornatari* did revolt, as they did in 1920, occupying uncultivated lands, they were violently put down by an alliance of large and small landowners, gang bosses, and local police, leaving dozens dead.

Today, the main crop is tomatoes rather than wheat, but the system of *caporali* and *giornatari* persists. The difference is that now the gang bosses are Polish, exploiting other Poles, hired in their own country and bused south across Europe, where they soon descend into a form of labor slavery from which there is no easy escape. In many ways, the modern version is worse, as journalist Alessandro Leogrande suggested. At least in the old system the bosses lived with the *giornatari*, in the same communities, and relied on their labor, which helped limit their excesses. There are no such checks on the modern bosses—foreigners hiring other foreigners—who themselves are transients. The Italian farm owners collaborate with them, turning a blind eye to abuses—financial, physical, and psychological— because of the pressure on them to keep costs to an absolute minimum. "Because nothing, but nothing, down here can stop the tomato harvest. Not the weather, not the rain, not the hail; so we can forget the law or the denunciations by day laborers," wrote Leogrande. Nonetheless, recent legal action against the system, beginning with accusations made by three brave Polish workers, has had some effect. Since 2008, mechanical harvesters have started to replace handpicking, a sure sign that new laws and increasing checks on illegal labor practices are starting to have an effect.

Tomato cultivation has been associated with other ills, too. In a country where water is a scarce and often unpredictable resource, agriculture is responsible for more water consumption than any other sector. Paradoxically, agriculture also is the main culprit for its pollution. The massive use of chemical fertilizers and herbicides end up in the water supply, posing a threat to human health. The Sarno, one of the rivers (along with the Sele) that enabled the success of tomato cultivation and production in the province of Salerno in the early twentieth century, is now its victim. Huge amounts of water are taken from the Sarno during the fifty or sixty days of the growing season. Actually, nobody seems to know how much water is used, or even how much is really needed. What does the river get in return? A reddish liquid, colored by the skins and seeds left over from canning. The river, which is only 15 miles long, is today one of Europe's most polluted.

New hybrid varieties of tomatoes have vastly increased productivity since the 1950s. At the same time, local or traditional varieties (*varietà tipiche*) are trumpeted or given protected status. In Vittoria (Sicily), local tomatoes like the 'Ferrisi' and the 'Pachino', grown where the soil is slightly saline, which gives them a mineral taste, are highly prized by restaurants as far away as Turin. Farther north in Albenga (Liguria), another furrowed variety, the 'Cuor di bue' (Oxheart), underwent a resurgence beginning in the 1980s, after languishing for decades when rounder, smoother varieties had been the fashion (figure 36). In the Sarno River valley, the 'San Marzano' has been given "denomination of protected origin" status. A variety originally developed for the export market has become a proud part of the region's identity and is once again finding its way into cans—even if it now takes a laboratory to determine whether the seeds are genuine 'San Marzano'.

These developments were the result of a slight shift in the EU's policy. In the 1970s, the EU's laws made it illegal to sell any cultivar not on the national list of each member country. Passed to ensure seed quality and "purity," the laws had the effect of favoring uniformity and large

Figure 36 The revival of traditional tomatoes: a crate of 'Cuor di bue'. (Photograph by F. Gioberti, courtesy of the Cooperativa Ortofrutticola, Albenga)

commercial growers at the expense of traditional varieties and smaller growers, since it was expensive to "list" a tomato type. Recently, some flexibility has been introduced into the system, and traditional cultivars have been promoted.

In the 1990s, one-sixth of the Italian population made their own tomato *conserva*. Competing with these are ready-made sauces and other products, which abound as never before and now claim more than one-third of the processed-tomato market. Much of Italians' food is still locally grown and raised, and what was previously seen as a sign of backwardness is now one of the key planks of the Slow Food movement (another Italian export worldwide). Italians also continue to prefer "naturalness" in their foods, and accordingly, their consumption of frozen foods is one-quarter that of the British and one-third that of the French or Germans. In addition, the Italians have turned the much-vaunted "Mediterranean diet" into an ideology, albeit respecting less in practice its components of a high intake of grains, legumes, vegetables, and fruit and a low intake of animal fats, at least compared with thirty years ago.

Finally is the force of tradition in food habits, which is particularly striking in the millions of Italians outside Italy, including their generations of descendants. They, too, continue to bottle tomatoes in the "traditional" way. The troubled teenage protagonist of Melina Marchetta's novel *Looking for Alibrandi* (1992) may dread her family's "tomato day" and the embarrassment that would result if any of her school friends should see her helping out. But as a Sicilian Australian, she comes around in the end, praising the "tradition that we'll never let go . . . because like religion, culture is nailed into you so deep you can't escape it."

In the postmodern world, "tradition" (or something like it) and the "constant flow" of goods demanded by supermarkets (and their consumers) go hand in hand. Italy is no different from North America or northern Europe. Fresh tomatoes are available year-round, grown under plastic during the winter months, in places like Vittoria (Sicily). A few growers even use the same, almost otherworldly, techniques first developed by the Dutch, in which tomato vines 50 feet long, in computer-controlled environments, are given precise doses of water, fertilizers, and pesticides. The irony is in growing 'Pachino' tomatoes, famous for the slightly saline taste the soil imparted, in coconut-fiber mats, their roots never touching the local soil. This is indeed a far cry from the day when local farmers protected their tender plants from the wind with pads of the prickly pear cactus.

Mass distribution has transformed the food chain. Today, a supermarket consumer in Vittoria, shopping for those famous local tomatoes anytime between October and April, may be offered something *grown* locally, but the tomatoes will have traveled more than 1,000 miles in the process. From the countryside around Vittoria, the tomatoes travel by truck to Catania, then by ship to Naples, and then again by truck to Fondi (Lazio), Italy's largest wholesale fruit and vegetable market, where the supermarket chain purchases and redistributes them and the tomatoes begin their "return" journey. Vittoria's growers are lucky to earn 90 cents a pound for these greenhouse tomatoes, but the supermarket charges around $2.60 a pound. Most of the difference goes to pay the costs of the distribution system. Organized crime also takes its cut, in what has proved a lucrative business, with the Mafia insisting on using its own transport and packing companies; threatening producers, buyers, and wholesalers; and even trafficking in the illegal immigrants employed in wholesale fruit and vegetable markets like Milan's Ortomercato. Seventy years ago, New York's "Artichoke king" could only dream of market infiltration on such a scale.

Let us conclude this book with the most important issue of all: taste. Do today's tomatoes taste different from those of the past, as Citati maintains? The short answer is yes. But there is a long answer too, and it is much more interesting. First, although this may seem obvious, taste is subjective and can vary widely from culture to culture. Moreover, tastes change over time. The tomato depicted in Federico Cesi's *Erbario miniato* (see figure 5) may look like some modern-day furrowed varieties, but we cannot assume that it tasted like them.

Today's food scientists can identify specific tastes. As Harold McGee has written, fruits and vegetables are a mixture of sweet, acid, and bitter tastes. The tomato's low sugar content (3%, which is low for a fruit) combines with a large amount of savory glutamic acid and aromatic sulfur compounds—more common in meat than in fruits—to produce a unique flavor. This balance of sweetness and acidity goes well with a wide range of foods. Because glutamic acid is the active ingredient in the flavor enhancer monosodium glutamate (MSG), it is no surprise that what is botanically a fruit is most often used as a vegetable.

Of course, there are many kinds of tomatoes. The tomatoes we eat are the result of hundreds of years of work by farmers and plant breeders to bring out those qualities considered desirable or health giving and to reduce those considered unpleasant or harmful. This is where we leave the objective view of the scientist for the subjective realm of our own present and past.

As an illustration of the tomato's subjective present, let us compare two very different European countries: Britain and Italy. Today, the tomato has different "taste zones." If we looked at a British seed catalog at random—say, Mr *Fothergill's A–Z of Vegetables*—we would find that of the thirty-three varieties of tomato seeds offered for sale in 2007, the descriptions of thirteen contain the word "sweet." Thus we have expressions like "beautifully sweet," "very sweet," "wonderfully sweet," and, just to vary things a bit, "very high sugar content." The varietals' names echo this: 'Nectar', 'Sweet Millions', and so on. If we chose an Italian seed catalog, however, the picture would be very different. Of the thirty-eight tomato varieties listed in the *Fratelli Ingegnoli catalogo guida* for 2007, the descriptions of only four highlight sweetness. The emphasis instead is on tomatoes with a "firm pulp," "strong flavor," and "deep red color."

This difference between the British and Italian catalogs can be found in the very different ways that tomatoes are consumed in the two countries. Even the word "fresh" has different meanings. A tomato bought "fresh" in Britain is consumed as a salad tomato, but in Italy, a tomato bought "fresh" is just as likely to be used as a cooking ingredient or for further processing. Different tomato varieties are consumed because different flavors and consistencies are required. On the one hand, salad tomatoes have a high water content and thin skins and are sweet. Processing, on the other hand, requires firm, almost "dry" tomatoes that are not overly sweet, since sweetness will be brought out in the processing. The Italian taste for tomatoes thus has been conditioned as much by a preference for vegetables served firm (as opposed to mushy) as by the dominance of the processing industry in determining which varieties of tomato are grown and marketed, especially over the past hundred years or so.

Now let us look at the tomato's subjective past. Food scientists find it logical that tomatoes are used in a variety of ways, especially as a savory condiment. But for historians, at least for me, there is nothing preordained in the tomato's history, nothing inevitable about *pasta al pomodoro* or any of the other many different ways in which tomatoes are prepared and consumed in Italy. Furthermore, there is nothing inevitable about the tomato's becoming Italy's dominant "vegetable fruit," to the point that the average Italian consumes around 198 pounds of fresh tomatoes and almost 66 pounds of processed tomatoes each year. There is nothing inevitable about Italy's becoming Europe's premier tomato nation and the world's second-largest exporter of processed tomatoes, and nothing inevitable about the "Italian" tomato's becoming an international success story, having crossed the Atlantic Ocean several times and reaching even more distant shores (figure 37). Instead, the tomato's history in Italy is the result of a series of chance developments.

The tomato's success in both Italy and elsewhere is due in no small part to its malleability. For two centuries, Italians limited the use of the few tomato varieties they had, and their early confusion with the toma-tillo points to the tartness of those tomatoes. Dietary rules and practices restricted the uses of a food with such qualities. But as medical ideas shifted, so did the place of the acidic tomato, which found an ever wider place as a condiment, stimulating the appetite and aiding digestion. By the end of the eighteenth century, growers were beginning to develop

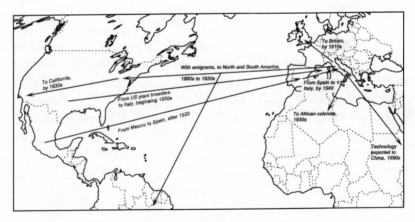

Figure 37 The well-traveled "Italian" tomato. (Map by the author)

new varieties, more delicate and less acidic in taste. This, in turn, encouraged a wider use of tomatoes in sauces and even as a dish on their own. It was not long before tomatoes were being used as a substitute for meat in sauces. In the nineteenth century, they began their career as salad tomatoes, first among the poor, and in the second half of the century, new, smaller varieties were used to season pizza and pasta. These varieties were especially well suited to being canned whole, whereas the larger varieties were turned into paste.

This is the second factor in the tomato's particular evolution in Italy: its close association with preservation and processing. The desire to consume tomatoes as a seasoning throughout the year resulted in their being pickled or sun dried, producing full-flavored condiments. Large tomatoes were best suited to sun drying, as well as to the means of preservation developed next, that of turning them into a paste. At first, this paste was dry and used like a spice. It was dark colored, had a very strong flavor, and was used sparingly. But with the invention of canning, tomato paste could be produced in a less concentrated, more liquid form, which retained the tomatoes' red color, and producers sought less and less acidic varieties.

At the same time, smaller, egg-shaped varieties were proving ideal for canning whole, with their skins removed. As tomato processing became an industry itself and an important export, new tomato varieties were developed—redder, sweeter, firmer, and more productive. Processing brought out their sweetness, with *pelati* becoming an acceptable substitute for fresh tomatoes. Indeed, Italians found that processed tomatoes were even better suited to sauces than were fresh tomatoes, and their association with industrially produced pasta was born.

This is where human ingenuity, our final factor in the evolution of the tomato in Italy, comes in, first in the form of seasoning, condiment, and sauce. Necessity played an important part, too, as poor Italians resorted to eating what little produce was available during the hot, dry summer months. Raw tomatoes offered a change from the "green" diet of vegetables that was the peasant's mainstay. The summer salad, in its various forms—the southern *acqua-sale*, the Tuscan *panzanella*, and the

Ligurian *condiglione*—became part of Italian regional cooking, especially as Italians gradually became less dependent on it in order to survive. In its cooked form, the tomato found a place in an abundance of dishes. Of course, the greatest leap forward was the creative association of tomato sauce with what was fast becoming a staple: pasta. More than anything else, more than even the EU's subsidies, the invention and popularity of *pasta al pomodoro* as a symbol of Italian cooking were behind the success of the Italian tomato worldwide.

Closer to our time, ingenuity in the form of technology has been important as well. Tomato concentrate appears in a range of processed foods, from canned sauces to frozen pizza, and has experienced a popular resurgence in response to claims about the health benefits of lycopene. Even though it may not yet be genetically modified, today's tomato is a medical "superfood." Even the plant's leaves—whose smell repulsed the first botanists and which have long been regarded as toxic because of their effects on pests—may be good for us, as the tomatine they contain apparently reduces cholesterol absorption. The chefs who have recently started to add a sprinkling of leaves in their tomato sauces toward the end of cooking, to impart a freshly picked tomato smell, may have discovered something new about the tomato.

Bibliography

Albala, Ken. *Eating Right in the Renaissance*. Berkeley: University of California Press, 2002.

Camporesi, Piero. *The Magic Harvest: Food, Folklore and Society*. Translated by Joan Krakover Hall. Cambridge: Polity Press, 1993.

Capatti, Alberto. "The Identity of the Tomato in Gastronomy" [in Italian]. *Rivista di antropologia* 76, suppl. (1998): 245–54.

Capatti, Alberto, A. De Bernardi, and A. Varni, eds. *Storia d'Italia. Annali 13, L'alimentazione*. Turin: Einaudi, 1998. (Hereafter abbreviated as *Annali 13*)

Capatti, Alberto, and Massimo Montanari. *La cucina italiana: Storia di una cultura*. Rome: Laterza, 1999.

———. *Italian Cuisine: A Cultural History*. Translated by Aine O'Healy. New York: Columbia University Press, 2003.

Carucci, Paola, ed. *Gli archivi per la storia dell'alimentazione*. 3 vols. Rome: Ministero per i beni culturali e ambientali, 1995. (Hereafter abbreviated as *Archivi dell'alimentazione*)

Cavaciocchi, Simonetta, ed. *Alimentazione e nutrizione, secc. XIII–XVIII: Atti della ventottesima settimana di studi*. Florence: Le Monnier, 1997.

Cesari Sartoni, Monica. *Mangia italiano: Guida alle specialità regionali italiane*. Milan: Morellini, 2005.

Chiappelli, Fredi, ed. *First Images of America: The Impact of the New World on the Old*. 2 vols. Berkeley: University of California Press, 1976.

Coats, Alice. "The Fruit with the Shady Past: Tales of the Tomato." *Country Life*, May 17, 1973.

Faccioli, Emilio, ed. *L'arte della cucina in Italia*. Turin: Einaudi, 1987.

Filippa, Marcella, ed. *Il cibo dell'altro: Movimenti migratori e culture alimentari nella Torino del Novecento.* Rome: Edizioni Lavoro, 2003. (Hereafter abbreviated as *Cibo dell'altro*)

Flandrin, Jean-Louis, and Massimo Montanari, eds. *Food: A Culinary History from Antiquity to the Present.* New York: Columbia University Press, 1999.

Goldman, Amy. *The Heirloom Tomato, from Garden to Table.* New York: Bloomsbury, 2008.

Grewe, Rudolph. "The Arrival of the Tomato in Spain and Italy: Early Recipes." *Journal of Gastronomy* 3 (1988): 67–81.

Harvey, Mark, Steve Quilley, and Huw Benyon. *Exploring the Tomato: Transformations of Nature, Society and Economy.* Cheltenham: Elgar, 2002.

Luckwill, Leonard. *The Genus* Lycopersicon: *An Historical, Biological, and Taxonomic Survey of the Wild and Cultivated Tomatoes.* Aberdeen: University Press, 1943.

McCue, George. "The History and Use of the Tomato: An Annotated Bibliography." *Annals of the Missouri Botanical Garden* 39 (1952): 289–348.

Ostuni, Nicola. "L'industria delle conserve di pomodoro in Campania." In *Las industrias agroalimentarias en Italia y España durante los siglos XIX y XX,* edited by Carlos Barciela López and Antonio Di Vittorio. Alicante: Universidad de Alicante, 2003.

Sentieri, Maurizio, and Guido Zazzu. *I semi dell'Eldorado: L'alimentazione in Europa dopo la scoperta dell'America.* Bari: Dedalo, 1992.

Smith, Andrew. *The Tomato in America: Early History, Culture, and Cookery.* Urbana: University of Illinois Press, 2001.

Preface

Saba, Umberto. "Bianco immacolato signore." 1946/1947. Quoted in Chiara Fabbian, "Poesia in cucina: Lina e il mondo femminile nella rappresentazione di Umberto Saba." *Modern Language Notes* 117 (2002): 174–93.

"Speciale pomodoro." *Agrisole,* suppl. 3, January 20–26, 2006.

1. "Strange and Horrible Things"

Ambrosoli, Mauro. *The Wild and the Sown: Botany and Agriculture in Western Europe, 1350–1850.* Translated by Mary McCann Salvatorelli. Cambridge: Cambridge University Press, 1997.

Castelvetro, Giacomo. *The Fruit, Herbs and Vegetables of Italy, an Offering to Lucy, Countess of Bedford.* Translated and edited by G. Riley. London: British Museum, 1989.

Cattabiani, Alfredo. *Florario: Miti, leggende e simboli di fiori e piante.* Milan: Mondadori, 1997.

Cesalpino, Andrea. *De plantis libri xvi.* Florence: Georgium Marescottum, 1583.

Cottino, Alberto, ed. *La natura morta al tempo di Caravaggio.* Naples: Electa, 1995.

Durante, Castore. *Herbario nuovo*. Rome: B. Bonfadino and T. Diani, 1585.

Eiche, Sabine. *Presenting the Turkey: The Fabulous Story of a Flamboyant and Flavourful Bird*. Florence: Centro Di, 2004.

Felici, Costanzo. *Costanzo Felici da Piobbico: Lettera sulle insalate, lectio nona de fungis*. Edited by Enzo Cecchini. Urbino: Arti grafiche editoriali, 1996.

Ferrières, Madeleine. *Sacred Cow, Mad Cow: A History of Food Fears*. Translated by Jody Gladding. New York: Columbia University Press, 2006.

Franchini, Dario, Renzo Margonari, Giuseppe Olmi, Rodolfo Signorini, Attilio Zanca, and Chiara Tellini Perina. *La scienza a corte: Collezionismo eclettico, natura e immagine a Mantova fra Rinascimento e Manierismo*. Rome: Bulzoni, 1979.

Freedberg, David. *The Eye of the Lynx: Galileo, His Friends, and the Beginning of Modern Natural History*. Chicago: University of Chicago Press, 2002.

Galluzzi, Riguccio. *Istoria del Granducato di Toscana sotto il governo della casa Medici*. Livorno: Vignozzi, 1820.

Geiger, Benno. *I dipinti ghiribizzosi di Giuseppe Arcimboldi*. Florence: Vallecchi, 1954.

Gerard, John. *The Herball or Generall Historie of Plantes*. London: Norton and Whitakers, 1633.

Greenblatt, Stephen. *Marvelous Possessions: The Wonder of the New World*. Chicago: University of Chicago Press, 1991.

Grieco, Allen. "The Social Politics of Pre-Linnaean Botanical Classification." *I Tatti Studies* 4 (1991): 131–49.

Hernández, Francisco. *Rerum medicarum Novae Hispaniae thesaurus seu plantarum, animalium, mineralium mexicanorum historia*. Rome: Vitale Mascardi, 1658. Facsimile ed., Rome: Istituto poligrafico e zecca dello stato, 1992.

Honour, Hugh. *The European Vision of America*. Cleveland: Cleveland Museum of Art, 1975.

Janick, Jules, and Giulia Caneva. "The First Images of Maize in Europe." *Maydica* 50 (2005): 71–80.

Kaufman, Thomas. "The Allegories and Their Meaning." In *The Arcimboldo Effect: Transformations of the Face from the Sixteenth to the Twentieth Century*, edited by Pontus Hulton, 89–108. London: Thames & Hudson, 1987.

Lippi, Lorenzo. *Il Malmantile racquistato*. Florence: Barbera, 1861.

Mattioli, Pietro Andrea. *Di Pedacio Dioscoride Anazarbeo libri cinque*. Venice: N. De Bascarini, 1544; Venice: Vincentium Valgrisium, 1554; Venice: Felice Valgrisi, 1595.

McTighe, Sheila. "Foods and the Body in Italian Genre Paintings, About 1580: Campi, Passarotti, Carracci." *Art Bulletin* 86 (2004): 301–23.

Medici Archive Project. Documentary Sources Database. Available at http://documents.medici.org.

Michiel, Pietro Antonio. *I cinque libri di piante*. Edited by Ettore De Toni. Venice: Carlo Ferrari, 1940.

Pavord, Anna. *The Naming of Names: The Search for Order in the World of Plants*. London: Bloomsbury, 2005.

Sahagún, Bernardino de. *Historia general de las cosas de Nueva España*. Edited by Eduardo Seler. 5 vols. Mexico City: Robredo, 1938.

Sala, Giovanni Domenico. *De alimentis*. 1628. In *Ars medica*. Padua: Franciscum Bolzetta, 1641.

Salerno, Luigi. *La natura morta in Italia / Still Life Painting in Italy, 1560–1905*. Rome: Bozzi, 1984.

Soderini, Giovanvettorio. *Della cultura degli orti e giardini . . . ora per la prima volta pubblicato*. Florence: Stamperia del Giglio, 1814.

Spike, John. *Italian Still Life Paintings from Three Centuries*. Florence: Centro Di, 1983.

Tongiorgi Tomasi, Lucia. "La scultura bronzea." In *Scultura a Pisa tra Quattro e Seicento*, edited by Roberto Paolo Ciardi, Claudio Casini, and Lucia Tongiorgi Tomasi, 319–75. Pisa: Cassa di Risparmio di Pisa, 1987.

Vasari, Giorgio. *Lives of the Artists*. Florence, 1550, 1568. English translation available at http://www.efn.org/~acd/vite/VasariLives.html.

Zeri, Federico. *Diari di lavoro 2*, 92–103. Turin: Einaudi, 1976.

2. Death by Vegetables

Albala, Ken. "Insensible Perspiration and Oily Vegetable Humor: An Eighteenth-Century Controversy over Vegetarianism." *Gastronomica: The Journal of Food and Culture* 2, no. 3 (2002): 29–36, and 3, no. 2 (2003): 2–5.

Ambrosoli, Mauro. "Orti, vigne, giardini: Condizione familiare e produzione per il mercato tra primo Cinquecento e metà Ottocento." *Bollettino storico bibliografico subalpino* 101 (2003): 41–93.

Castelvetro, Giacomo. *The Fruit, Herbs and Vegetables of Italy*. Translated and edited by G. Riley. London: British Museum, 1989.

Ciuffoletti, Zeffiro, and Giuliano Pinto, eds. *Desinari nostrali: Storia dell'alimentazione a Firenze e in Toscana*. Florence: Polistampa, 2005.

Cocchi, Antonio. *The Pythagorean Diet, of Vegetables Only, Conducive to the Preservation of Health, and the Cure of Diseases*. London: R. Dodsley, 1745.

Dallington, Robert. *A Survey of the Great Dukes State of Tuscany, in the Yeare of Our Lord 1596*. London: Edward Blount, 1605.

D'Ambrosio, Angelo, and Mario Spedicato. *Cibo e clausura: Regimi alimentari e patrimoni monastici nel Mezzogiorno moderno (secc. XVII–XIX)*. Bari: Cacucci, 1998.

Demarco, Domenico, ed. *La "Statistica" del regno di Napoli nel 1811*. 4 vols. Rome: Accademia Nazionale dei Lincei, 1988.

Gallo, Agostino. *Le vinti giornate dell'agricoltura e dei piaceri della villa.* Edited by Luciana Crosato Larcher. Treviso: Canova, 2003.

Gerard, John. *The Herball or Generall Historie of Plantes.* London: Norton and Whitakers, 1633.

Goethe, Johann Wolfgang von. *Italian Journey, 1786–1788.* Translated by W. H. Auden and Elizabeth Mayer. Harmondsworth: Penguin, 1982.

Gratarolo, Guglielmo. *A Direction for the Health of Magistrates and Studentes.* London: William How, 1574.

Guerrini, Anita. "A Diet for the Sensitive Soul: Vegetarianism in Eighteenth-Century Britain." *Eighteenth-Century Life* 23 (1999): 34–42.

Jacquot, Félix. *Lettres médicales sur l'Italie, comprenant l'histoire médicale du coprs d'occupation des États romains.* Paris: Masson, 1857.

Malacarne, Giancarlo. *Sulla mensa del principe: Alimentazione e banchetti alla corte dei Gonzaga.* Modena: Il Bulino, 2001.

Martino of Como. *The Art of Cooking: The First Modern Cookery Book.* Edited by Luigi Ballerini. Berkeley: University of California Press, 2005.

Massonio, Salvatore. *Archidipno, ovvero dell'insalata e dell'uso di essa.* Venice: Marc'Antonio Brogiollo, 1628.

Milazzo, Giuseppe. *Piazza delle erbe: La storia dei Ciassè e dell'antico Palazzo di Giustizia di Savona.* Savona: Daner, 1998.

Palmer, Richard. "Health, Hygiene and Longevity in Medieval and Renaissance Europe." In *History of Hygiene,* edited by Yosio Kawakita, Shizu Sakai, and Yasuo Otsuka, 75–98. Tokyo: Ishiyaku EuroAmerica, 1991.

Pelling, Margaret. "Food, Status and Knowledge: Attitudes to Diet in Early Modern England." In *The Common Lot: Sickness, Medical Occupations and the Urban Poor in Early Modern England,* edited by Margaret Pelling, 38–62. London: Longman, 1998.

Pujati, Giovanni Antonio. *Riflessioni sul vitto pitagorico.* Feltre: Odoardo Foglietta, 1751.

Rangone, Tommaso. *De vita hominis ultra cxx annos potrahenda.* Venice: Nicolinus Sabiensis, 1550.

Romoli, Domenico. *La singolar dottrina di m. D. Romoli detto il Panonto.* Venice: G. B. Combi, 1637.

Sacchi, Bartolomeo. *Platina: On Right Pleasure and Good Health.* Edited and translated by Mary Ella Milham. Tempe: Arizona State University Press, 1998.

Sada, Luigi. "L'arte culinaria barese al celebre banchetto nuziale di Bona Sforza nel 1517." In *La regina Bona Sforza tra Puglia e Polonia,* edited by Bronisław Biliński, 41–61. Warsaw: Ossolineum, 1987.

Sereni, Emilio. "Note di storia dell' alimentazione del Mezzogiorno: I napoletani da 'mangiafoglia' a 'mangiamaccheroni.'" *Cronache meridionali* 5 (1958): 272–95, 353–77.

Thirsk, Joan. *Food in Early Modern England: Phases, Fads, Fashions, 1500–1760.* London: Hambledon Continuum, 2007.

Toaff, Ariel. *Mangiare alla giudia: La cucina ebraica in Italia dal Rinascimento all'età moderna.* Bologna: Il Mulino, 2000.

3. "They Are to Be Enjoyed"

Agnoletti, Vincenzo. *Manuale del cuoco e del pasticciere.* Pesaro: Nobili, 1832.

Anonimo Piemontese. *Descrizione dell'isola di Sardegna.* Edited by Francesco Manconi. Cagliari: Comune di Cagliari, 1985.

Baldissarro, Lia Domenica. "Osservazioni sull'alimentazione nel secolo XVIII attraverso le carte di amministrazione di due famiglie nobili: Milano-Franco di Polistena e Saluzzo di Corigliano." In *Archivi dell'alimentazione* 2:1250–65.

Belli, Giuseppe Gioacchino. *Le lettere.* Edited by Giacinto Spagnoletti. Milan: Cino del Duca, 1961.

Bicchierai, Luigi. *Pennino l'oste.* Edited by Franco Tozzi. Signa: Masso delle Fate, 1996.

Castelli, Pietro. *Antidotario romano commentato dal Dottor Pietro Castelli.* Messina: Vedova di Gio. Francesco Bianco, 1637.

———. *Hortus messanensis.* Messina: Branco, 1640.

Coles, Elisha. *An English Dictionary Explaining the Difficult Terms That Are Used in Divinity, Husbandry, Physick, Philosophy, Law, Navigation, Mathematicks, and Other Arts and Sciences.* London: Peter Parker, 1677.

Corrado, Vincenzo. *Del cibo pitagorico ovvero erbaceo, per uso de' nobili e de' letterati.* Naples, 1781. Reprint, Rome: Donzelli, 2001.

———. *Il cuoco galante, opera meccanica.* 1773. Naples: Raimondi, 1786. Facsimile ed., Sala Bolognese: Forni, 1990.

D'Ambrosio, Angelo, and M. Spedicato. *Cibo e clausura: Regimi alimentari e patrimoni monastici nel Mezzogiorno moderno (sec. XVII–XIX).* Bari: Cacucci, 1998.

Demarco, Domenico, ed. *La "Statistica" del regno di Napoli nel 1811.* 4 vols. Rome: Accademia Nazionale dei Lincei, 1988.

Fondo Gesuitico: "Entrata ed uscita del Collegio Romano di più decenni," b. 1363:1. Biblioteca Nazionale Centrale, Rome.

Fongaro, Maria Carmela. "Alimentazione di prelati, monache e . . . fattori a Brindisi nel secoli XVII e XVIII." In *Archivi dell'alimentazione* 2:1529–42.

Gaudentio, Francesco. "Il Panunto toscano, overo la teologia gaudentiana, opera in cui si mostra il modo facile di cocinar moderno con poca spesa . . . scritto in Roma l'anno del Signore 1705 per uso dell'infermaria del Collegio Romano," MS 450. Biblioteca Città di Arezzo.

———. *Il panunto toscano ritrovato e commentato.* Edited by G. Gianni. Rome: Trevi, 1976.

Guigoni, Alessandra. "L'acculturazione agroalimentare delle piante americane in Sardegna." Ph.D. diss., Università degli Studi di Siena, 2003.

———. "Rassegna delle fonti storiche sull'introduzione di alcune piante americane in Sardegna." *Quaderni Bolatanesi* 31 (2005): 245–66.

Latini, Antonio. *Autobiografia (1642–96): La vita di uno scalco.* Edited by F. Luccichenti. Rome: Leberit, 1992.

———. *Lo scalco alla moderna.* 2 vols. Naples: Domenico Antonio Parrino e Michele Mutii, 1692, 1694.

Mariani, Luca, Agata Parisella, and Giovanna Trapani. *La pittura in cucina.* Palermo: Sellerio, 2003.

Miller, Philip. *The Gardener's Dictionary: Containing the Best and Newest Methods of Cultivating and Improving the Kitchen, Fruit, Flower Garden and Nursery.* 8th ed. London: Printed for the author, 1768.

Pedrocco, Giorgio. "La conservazione del cibo: Dal sale all'industria agro-alimentare." In *Annali 13,* 379–447.

Provincia Romana: "Levitico della Provincia," b. 199; "Diario della Casa Professa di Roma, 1757–63," b. 211; and "Loreto Coll. Il lirico, memoriale, 1720–3, b. 216." Archivium Romanum Societatis Iesu (Archives of the Society of Jesus), Rome.

Ray, John. *Historia plantarum.* 2 vols. London: Clark, 1686, 1688.

———. *Travels Through the Low-Countries, Germany, Italy and France.* 2 vols. London, 1730.

Re, Filippo. *L'ortolano dirozzato.* Milan: Giovanni Silvestri, 1811.

Rizzo Pavone, Renata, and Anna Maria Iozzia. "Comunità monastiche catanesi tra '700 e '800." In *Archivi dell'alimentazione* 3:1835–51.

Sereni, Emilio. *History of the Italian Agricultural Landscape.* Translated by R. Burr Litchfield. Princeton, N.J.: Princeton University Press, 1997.

Spatuzzi, Achille, Luigi Somma, and Errico De Renzi. *Sull'alimentazione del popolo minuto in Napoli: Lavori due approvati dall'Accademia Pontaniana.* Naples: Stamperia della R. Università, 1863.

Targioni Tozzetti, Giovanni. *Ragionamenti . . . sull'agricoltura Toscana.* Lucca: Jacopo Giusti, 1759.

Targioni Tozzetti, Ottaviano. *Istituzioni botaniche.* Vol. 2. Florence, 1813.

Tino, Pietro. "Napoli e i suoi dintorini: Consumi alimentari e sistemi colturali nell'Ottocento." *Meridiana* 18 (1993): 47–99.

Valdès, Marina. "L'aristocrazia cagliaritana a tavola: Cultura e abitudini alimentari fra '700 e '800." In *Archivi dell'alimentazione* 3:1964–2016.

Zampetti, Pietro, ed. *Carlo Magini.* Milan: Motta, 1990.

Zannoni, Mario. *A tavola con Maria Luigia: Il servizio di bocca della duchessa di Parma dal 1815 al 1847.* Parma: Artegrafica Siera, 1991.

4. Pasta al Pomodoro

Artusi, Pellegrino. *La scienza in cucina e l'arte di mangiar bene*. 1891. Edited by Piero Camporesi. Turin: Einaudi, 1970.

————. *Science in the Kitchen and the Art of Eating Well*. Translated by Murtha Baca and Stephen Sartarelli. Toronto: University of Toronto Press, 2003.

Basile, Giambattista. *Lo cunti de li cunti*. Edited by M. Rak. Milan: Garzanti, 1986.

Bazzi, Andreina. "Il ricettario di Lucia Prinetti Adamoli." In *Archivi dell'alimentazione* 2:1266–82.

Benincasa, Gabriele. *La pizza napoletana: Mito, storia e poesia*. Naples: Guida, 1992.

Biasin, Gian-Paolo. *The Flavors of Modernity: Food and the Novel*. Princeton, N.J.: Princeton University Press, 1993.

Bicchierai, Luigi. *Pennino l'oste*. Edited by Franco Tozzi. Signa: Masso delle Fate, 1996.

Bonsignori, Giovanni. *La coltivazione del pomodoro*. Brescia: Queriniana, 1901.

Bourcard, Francesco de, ed. *Usi e costumi di Napoli e contorni, descritti e dipinti*. 2 vols. Naples: G. Nobile, 1853–1858. [Includes Carlo Tito Dal Bono, "Le taverne"]

Canevazzi, Eugenio, and Francesco Marconi. *Vocabolario di agricoltura*. 2 vols. Rocca San Casciano: Cappelli, 1892.

Catalogo di semi e piante disponibili nell'Orto sperimentale della Società economica di Terra di Lavoro, per la primavera e l'autunno del 1861. Naples: Fibreno, 1861.

Cavalcanti, Ippolito. "Cucina casarinola co la lengua napolitana." Appendix, in Neapolitan, to *Cucina teorico-pratica*. Naples: Luigi Marotta, 1837.

Collodi, Carlo [Carlo Lorenzini]. *The Adventures of Pinocchio*. Translated by M. A. Murray. New York: Grosset & Dunlap, 1965.

————. *Il viaggio per l'Italia di Giannettino*. Florence: Paggi, 1882–1886.

La cucina casereccia: Per istruzione di chi prende diletto di fornire di vivande una mensa accoppiando al gusto l'economia. Naples: Giordano, 1816.

Demarco, Domenico, ed. *La "Statistica" del regno di Napoli nel 1811*. 4 vols. Rome: Accademia Nazionale dei Lincei, 1988.

Dumas, Alexandre. *Impressions de voyage: Le Corricolo*. Paris, 1841–1843.

Felici, Costanzo. *Costanzo Felici da Piobbico: Lettera sulle insalate, lectio nona de fungis*. Edited by Enzo Cecchini. Urbino: Arti grafiche editoriali, 1996.

Gallino, Giacinto. *Le barufe in famegia*. Milan: Treves, 1872.

Gasparrini, Guglielmo. *Cenno dello stato presente dell'agricoltura della provincia di Napoli*. In *Breve ragguaglio dell'agricoltura e pastorizia del regno di Napoli di qua del Faro*. Naples: Filiatre, 1845.

Goethe, Johann Wolfgang von. *Italian Journey, 1786–1788*. Translated by W. H. Auden and Elizabeth Mayer. Harmondsworth: Penguin, 1982.

Grimod de la Reynière, Alexandre-Balthazar-Laurent. *L'Almanach des gourmands.* 8 vols. Paris: Chez Maradan, 1803–1812.

La Cecla, Franco. *La pasta e la pizza.* Bologna: Il Mulino, 1998.

Lalande, Jérôme de. *Voyage d'un français en Italie, fait dans les années 1765–1766.* Venice, 1769.

Lenormant, François. *La grand Grèce: Paysages et histoires.* 3 vols. Paris: A. Lévy, 1881–1884.

Levi, Carlo. *Christ Stopped at Eboli: The Story of a Year.* Translated by Frances Frenaye. New York: Farrar, Straus, 1947.

Mutti, Ugo. "Carlo Rognoni: Un pioniere dell'agricoltura parmense." *Archivio storico per le provincie parmensi* 35 (1983): 239–48.

Perugini Antonini, Giuseppina. *Mangiar friulano.* Venice: Pozza, 1963.

Prezzolini, Giuseppe. *A History of Spaghetti Eating and Cooking For: Spaghetti Dinner.* New York: Abelard-Schuman, 1955.

Ray, John. "Several Sorts of Meats, Fruits, Sallets, Etc. Used in Italy and Other Observations on Diet." In *Travels Through the Low-Countries, Germany, Italy and France,* 1:346–50. London, 1730.

Roda, Marcellino, and Giuseppe Roda. *Manuale dell'ortolano contenente la coltivazione ordinaria e forzata delle piante d'ortaggio.* Turin: Unione tipografico-editrice, 1868.

Salini, Andrea. *Educare al lavoro: L'Istituto Artigianelli di Brescia e la colonia agricola di Remedello Sopra tra '800 e '900.* Milan: Angeli, 2005.

Scafoglio, Domenico, and Luigi Lombardi Satriani. *Pulcinella: Il mito e la storia.* Milan: Leonardo, 1990.

Serao, Matilde. *Il ventre di Napoli.* Milan: Treves, 1884.

———. *Racconti napoletani.* Milan: Treves, 1889.

Sereni, Emilio. "I Napoletani, da 'mangiafoglia' a 'mangiamaccheroni.'" *Cronache meridionali* 5 (1958): 272–95, 353–77.

Serventi, Silvano, and Françoise Sabban. *Pasta: The Story of a Universal Food.* Translated by Antony Shugaar. New York: Columbia University Press, 2002.

Smith, Andrew. *Pure Ketchup: A History of America's National Condiment.* Washington, D.C.: Smithsonian Institution Press, 2001.

———. *The Tomato in America: Early History, Culture, and Cookery.* Urbana: University of Illinois Press, 2001.

Snowden, Frank. *The Conquest of Malaria: Italy, 1900–1962.* New Haven, Conn.: Yale University Press, 2006.

Somogyi, Stefano. "L'alimentazione nell'Italia unita." In *Storia d'Italia,* vol. 5, no. 1, *I documenti,* edited by R. Romano, 841–87. Turin: Einaudi, 1973.

Spatuzzi, Achille, Luigi Somma, and Errico De Renzi. *Sull'alimentazione del popolo minuto in Napoli: Lavori due approvati dall'Accademia Pontaniana.* Naples: Stamperia della R. Università, 1863.

Teti, Vito. "Le culture alimentari nel Mezzogiorno continentale in età contemporanea." In *Annali 13*, 65–165.

———. *Storia del peperoncino: Un protagonista delle culture mediterranee*. Rome: Donzelli, 2007.

Tino, Pietro. "Napoli e i suoi dintorini: Consumi alimentari e sistemi colturali nell'Ottocento." *Meridiana* 18 (1993): 47–99.

Vamba [Luigi Bertelli]. *Il giornalino di Gian Burrasca*. Florence: Giunti, 1919.

Verga, Giovanni. *I Malavoglia*. 1881. Available at www.liberliber.it/biblioteca.

———. *The House by the Medlar Tree*. Translated by Eric Mosbacher. New York: Grove Press, 1953. [The only English translation I had available at the time of writing, this is abridged and lacks relevant passages, so I have supplied my own translation]

Vilmorin-Andrieux and Company. *The Vegetable Garden*. Translated by W. Miller. London: Murray, 1885.

Vivenza, Andrea. *Monografia sulla coltivazione del pomodoro in Abruzzo e sulla preparazione dell'estratto o conserva*. Piacenza: Solari, 1879.

5. "Authentic Italian Gravy"

Baldoni, Remigio. *Il pomodoro industriale e da tavola*. Rome: Ramo editoriale degli agricoltori, 1940.

Barr, Nancy Verde. *We Called It Macaroni: An American Heritage of Southern Italian Cooking*. New York: Knopf, 1992.

Bevilacqua, Piero. "Emigrazione transoceanica e mutamenti dell'alimentazione contadina calabrese fra Ottocento e Novecento." *Quaderni storici*, no. 47 (1981): 520–55.

Blaine, Julian, Angelo Calabrese, and Luigi Castellano. *Pianeta rosso: Arte e industria conserviera*. Naples: Pacifico, 1999.

Breckinridge, Sophonisba. P. *New Homes for Old*. New York: Harper, 1921.

Canadé Sautman, Francesca. "Women of the Shadows: Italian American Women, Ethnicity, and Racism in American Cinema." *Differentia*, nos. 6–7 (1994): 219–46.

Ciambelli, Bernardino. *I misteri di Mulberry*. [New York]: Frugone & Balletto, 1893.

Cinotto, Simone. *Una famiglia che mangia insieme: Cibo ed etnicità nella comunità italo-americana di New York, 1920–1940*. Turin: OTTO, 2001.

Corti, Paola. "Emigrazione e consuetudini alimentari: L'esperienza di una catena migratoria." In *Annali 13*, 683–719.

Covello, Leonard. Leonard Covello Papers. Balch Institute for Ethnic Studies, Philadelphia, Pa.

De Amicis, Edmondo. *Sull'oceano*. 1889. Available at http://www.liberliber.it/biblioteca.

DeAngelis, Rose, and Donald Anderson. "Gastronomic *Miscuglio*: Foodways in Italian-American Narrative." *Italian Americana* 23 (2005): 48–68.

DeLillo, Don. *Underworld*. New York: Scribner, 1997.

Durante, Francesco. *ItaloAmericana: Storia e letteratura degli Italiani negli Stati Uniti, 1880–1943*. Milan: Mondadori, 2005.

Fante, John. "A Wife for Dino Rossi." 1940. In *The Wine of Youth: Selected Stories*, 71–110. New York: HarperCollins, 1985.

Foerster, R. F. *The Italian Emigration of Our Times*. Cambridge, Mass.: Harvard University Press, 1919.

Gabaccia, Donna. "Italian-American Cookbooks: From Oral to Print Culture." *Italian Americana* 16 (1998): 15–23.

———. *Italy's Many Diasporas*. London: UCL Press, 2000.

———. *We Are What We Eat: Ethnic Food and the Making of Americans*. Cambridge, Mass.: Harvard University Press, 1998.

Gambino, Richard. *Blood of My Blood: The Dilemma of the Italian-Americans*. Garden City, N.Y.: Anchor Press, 1974.

Gentile, Maria. *The Italian Cookbook*. New York: Italian Book Company, 1919. Also available at http://digital.lib.msu.edu/projects/cookbooks/html/books/book_71.cfm.

Harney, Robert. *Italians in Canada*. Occasional Papers on Ethnic and Immigration Studies, no. 1. Toronto: Multicultural History Society of Ontario, 1978.

La Sorte, Michael. *La Merica: Images of Italian Greenhorn Experience*. Philadelphia: Temple University Press, 1985.

Laurino, Maria. "Words." In *The Milk of Almonds: Italian American Women Writers on Food and Culture*, edited by Louise DeSalvo and Edvige Giunta, 312–20. New York: Feminist Press at the City University of New York, 2002.

Levenstein, Harvey. "The American Response to Italian Food, 1880–1930." *Food and Foodways* 1 (1985): 1–24.

Mangione, Jerre. *Mount Allegro: A Memoir of Italian American Life*. 1943. Reprint, New York: Columbia University Press, 1981.

Minami, K. "The History of Tomato Production for Industry in South America." *Acta Horticulturae* 100 (1980): 87–92.

Murray, Catherine Tripalin. *A Taste of Memories from the Old "Bush."* Madison, Wis.: Italian-American Women's Mutual Society, 1988.

Orsi, Robert A. "The Fault of Memory: 'Southern Italy' in the Imagination of Immigrants and the Lives of Their Children in Italian Harlem, 1920–1945." *Journal of Family History* 15 (1990): 133–47.

Ortoleva, Peppino. "La tradizione e l'abbondanza: Riflessioni sulla cucina degli Italiani d'America." *Altreitalie* 7 (1992): 12–31. Also available at http://www.altreitalie.it/.

Prezzolini, Giuseppe. *A History of Spaghetti Eating and Cooking For: Spaghetti Dinner.* New York: Abelard-Schuman, 1955.

Puzo, Mario. *The Fortunate Pilgrim.* New York: Lancer, 1964.

Rovetta, Renato. *Il pomodoro: Coltivazione, industria, fabbricazione delle scatolette di latta.* Milan: Hoepli, 1914.

Vergara, Joe. *Love and Pasta: A Recollection.* New York: Harper & Row, 1968.

Zago, Ferruccio. *La coltivazione industriale del pomodoro.* Casale: Ottavi, 1913.

6. The Autarchical Tomato

Adorno, Salvatore. *Gli agrari a Parma: Politica, interessi e conflitti di una borghesia padana in età giolittiana.* Reggio Emilia: Diabasis, 2007.

Agnetti, Vittorio. *La nuova cucina delle specialità regionali.* Milan: Società editoriale Milanese, 1909.

Alliata di Salaparuta, Enrico. *Cucina vegetariana: Manuale di gastrosofia naturista.* Milan: Hoepli, 1930.

Arvidsson, Adam. "Between Fascism and the American Dream: Advertising in Interwar Italy." *Social Science History* 25 (2000): 151–86.

Baldoni, Remigio. *Il pomodoro industriale e da tavola.* Rome: Ramo editoriale degli agricoltori, 1940.

Cirio. *Nuovi orizzonti per la vostra mensa.* Preface by Lidia Morelli. Portici, 1936.

——. *Pomodoro fresco a ogni stagione.* Preface by Lidia Morelli. San Giovanni a Teduccio: Cirio, 1939.

Colapinto, Leonardo, and Antonino Annetta. *Il farmaco nel periodo autarchico.* Sansepolcro: Aboca, 2005.

Helstosky, Carol. *Garlic and Oil: Politics and Food in Italy.* Oxford: Berg, 2004.

Levi, Carlo. *Christ Stopped at Eboli: The Story of a Year.* Translated by Frances Frenaye. New York: Farrar, Straus, 1947.

Longarini, Pier Luigi. *Il passato . . . del pomodoro: La storia delle fabbriche di salsa nel parmense attraverso le cartoline, le scatole, i manifesti, le fotografie, i documenti.* Parma: Silvia, 1998.

Marinetti, Filippo Tommaso, and Fillìa [Luigi Colombo]. *La cucina futurista.* 1932. Reprint, Milan: Marinotti, 1998.

Marvin, Thomas. *Canned Tomatoes and Tomato Paste: Report of the United States Tariff Commission to the President of the United States.* Washington, D.C.: Government Printing Office, 1929.

Mauri, Giovanni. "Il sestuplo concentrato di pomodoro." In *Atti del primo convegno scientifico-tecnico per le conserve alimentari*. Parma, 1942.

Moretti Foggia, Amalia [Petronilla, pseud.]. *Desinaretti per . . . questi tempi*. Milan: Sonzogno, 1944.

————. *Ricette di Petronilla per tempi eccezionali*. Milan: Sonzogno, 1941.

————. *200 suggerimenti per . . . questi tempi*. Milan: Sonzogno, 1943.

Moroni Salvatori, Maria Paola. "Ragguaglio bibliografico sui ricettari del primo Novecento." In *Annali 13*, 887–925.

Norman, Jill, ed. *Eating for Victory: Healthy Home Front Cooking on War Rations*. London: O'Mara, 2007.

Peduto, Stefania. "Il settore conserviero dell'industria alimentare italiana tra proprietà pubblica e proprietà private: La storia della Cirio." Ph.D. diss., Università degli Studi di Salerno, 2003.

Pergreffi, Iacopo. *L'industria del pomodoro a Parma tra la fine dell'Ottocento e la seconda guerra mondiale*. Reggio Emilia: Tecnograf, 1994.

Randi, Elisabetta. *La cucina autarchica*. Florence: Cionini, 1942.

Rovetta, Renato. *Il pomodoro: Coltivazione, industria, fabbricazione delle scatolette di latta*. Milan: Hoepli, 1914.

Scicolone, Maria. *A tavola con il Duce: Ricette e racconti inediti di casa Mussolini*. Rome: Gremese, 2004.

Shriver, Alexis. *Canned-Tomato Industry in Italy*. Washington, D.C.: Government Printing Office, 1915.

Sorcinelli, Paolo. *Gli italiani e il cibo: Dalla polenta ai cracker*. Milan: Mondadori, 1999.

Teti, Vito. "Le culture alimentari nel Mezzogiorno continentale in età contemporanea." In *Annali 13*, 65–165.

Touring Club Italiano. *Guida gastronomica d'Italia*. Milan: Touring Club Italiano, 1931.

Trova, Assunta. "L'approvigionamento alimentare dell'esercito italiano dall'Unità alla seconda guerra mondiale." In *Annali 13*, 495–530.

L'Ufficio propaganda del Partito nazionale fascista. *Come si mangia in Europa*. Milan: Vitagliano, 1941.

Venè, Gian Franco. *Mille lire al mese: Vita quotidiana della famiglia nell'Italia fascista*. Milan: Mondadori, 1988.

Willson, Perry. "Cooking the Patriotic Omelette: Women and the Italian Fascist Ruralization Campaign." *European History Quarterly* 27 (1997): 531–47.

Zamagni, Vera. "L'evoluzione dei consumi fra tradizione e innovazione." In *Annali 13*, 171–204.

7. The Tomato Conquest

Anastasia, Mara, and Bruno Maida. "I luoghi dello scambio: Immigrazione e commercio alimentare a Torino nel secondo dopoguerra. In *Cibo dell'altro*, 3–52.

Bertagnoni, Giuliana. "Cibo e lavoro: Una storia della ristorazione aziendale in Italia." *Storia e futuro: Rivista di storia e storiografia* 15 (2007). Also available at http://www.storiaefuturo.com/articoli.

Beyers, Leen. "Creating Home: Food, Ethnicity and Gender Among Italians in Belgium Since 1946." *Food, Culture, and Society* 11, no. 1 (2008): 7–27.

Counihan, Carole. *Around the Tuscan Table: Food, Family, and Gender in Twentieth-Century Florence*. New York: Routledge, 2004.

David, Elizabeth. *Italian Food*. 1954. Reprint, Harmondsworth: Penguin, 1963.

Fiorini, Laura. "Salute e alimentazione durante gli anni Sessanta e Settanta a Torino nella memoria dei medici di famiglia." In *Cibo dell'altro*, 283–324.

Fissore, Gianpaolo. "Gli italiani e il cibo sul grande schermo dal secondo dopoguerra a oggi: Mito della convivialità e demarcazione delle differenze." In *Cibo dell'altro*, 163–79.

Ginsborg, Paul. *A History of Contemporary Italy: Society and Politics, 1943–1988*. London: Penguin, 1990.

Giordano, Arturo. *Pomodoro*. Rome: Istituto Poligrafico dello Stato, 1961.

Kinsella, M. N. "Tomato Industry in Australia." *Acta Horticulturae* 100 (1980): 63–69.

Lafranceschina, Felice. "Lucani a Torino: Storio di vita." Available at http://www.consiglio.basilicata.it/basilicata_regione_notizie/lucani_nel_mondo/.

Loren, Sophia. *In cucina con amore*. Milan: Rizzoli, 1971.

——. *Eat with Me*. London: Joseph, 1972.

Ludergnani, R. "Tomato Processing in Italy and Yugoslavia." *Acta Horticulturae* 100 (1980): 31–36.

Magelli, Ettore. *Coltivazione del pomodoro*. Florence: Vallecchi, 1954.

Maida, Bruno. "'Morti di fame': La rappresentazione della povertà sui quotidiani torinesi (1958–63)." In *Cibo dell'altro*, 203–38.

Margotti, Marta. "Il cibo degli operai: Trasformazioni sociali e culture alimentari a Torino dal 1945 al 1980." In *Cibo dell' altro*, 103–62.

Marotta, Giuseppe. *L'oro di Napoli*. Milan: Bompiani, 1947.

Mestdag, Inge. "Introducing Italian Cuisine into Flemish Home-Meal Cooking in the Twentieth Century: An Analysis of the Flemish Cook Bible *Ons kookboek* (1927–1999)." *Food and History* 1 (2003): 156–78.

Piovene, Guido. *Viaggio in Italia*. 1957. Reprint, Milan: Baldini & Castoldi, 1993.

Quarantotti De Filippo, Isabella. *Si cucine cumme vogli'i': La cucina povera di Eduardo De Filippo raccontata dalla moglie Isabella*. Milan: Tommasi, 2005.

Rea, Domenico. "I mesi." 1988. In *Domenico Rea: Opere*, edited by Francesco Durante, 1473–89. Milan: Mondadori, 2005.

Root, Waverley. *The Food of Italy*. New York: Atheneum, 1971.

Rossi Lodomez, Vera, and Franca Matricardi. *Il cucchiaio d'argento: Il libro fondamentale della cucina italiana*. Milan: Domus, 1950.

Savinio, Alberto. *Palchetti romani*. Milan: Adelphi, 1982.

Tasca, Luisa. "The 'Average Housewife' in Post–World War II Italy." *Journal of Women's History* 16 (2004): 92–115.

Vercelloni, Luca. "La modernità alimentare." In *Annali* 13, 949–1005.

Vittorini, Elio. *Conversazione in Sicilia*. Florence: Bompiani, 1941. [Originally published as *Nome e lagrime*]

———. *Conversation in Sicily*. Translated by Wilfrid David. London: Drummond, 1949.

Epilogue

Andrioli, Alberto. "I pomodori dei bei tempi andati." *L'Informatore Agrario*, September 2006, 1–7.

Biondani, Paolo. "La mafia è servita. Estorsioni, lavoro nero, tangenti: Così la criminalità controlla l'agroalimentare." *L'espresso*, May 28, 2009, 54–60.

Bunte, Frank, and Pim Roza. *Peeling Tomato Paste Subsidies: The Impact of a Revision of the CMO for Processing Tomatoes on European Horticulture*. Report 6.07.01. The Hague: LEI, 2007.

Cannata, Giuliano, and Maddalena Scaramella. *Acqua in Campania (e nel mondo)*. Naples: Guida, 2008.

Citati, Pietro. "Quando i pomodori avevano un sapore." *La Repubblica*, August 18, 2006, 1, 20.

Fratelli Ingegnoli catalogo guida 2007. Milan: Fratelli Ingegnoli, 2007.

Gatti, Fabrizio. "Io schiavo in Puglia." *L'espresso*, September 7, 2006, 28–35. English translation available at http://espresso.repubblica.it/dettaglio/.

Giuberti, Mirella, "Cenni storici e botanici, aspetti tecnologici e caratteristiche nutrizionali." Issue "Speciale pomodoro," *Zafferano Magazine*, July 2004. Also available at http://www.zafferano.org/rivista/17/pdf/star_bene.pdf.

Granello, Licia. "Dieta mediterranea addio: Così cambia la tavola italiana." *La Repubblica*, October 17, 2004, 28.

Jadeluca, Paola. "Riprendiamoci la terra, asset strategico della nuova economia." Interview with Carlo Petrini. *La Repubblica, Affari e finanza* sec., November 6, 2006.

Leogrande, Alessandro. *Uomini e caporali: Viaggio tra i nuovi schiavi nelle campagne del Sud*. Milan: Mondadori, 2008.

Manzoni Pubblicità. "Cirio e Depardieu 'tengono' un cuore italiano: Notizie dalle Aziende." *La Repubblica*, September 14, 2008, 26.

Marchetta, Melina. *Looking for Alibrandi.* Victoria: Puffin, 1992.

McGee, Harold. *On Food and Cooking: The Science and Lure of the Kitchen.* New York: Scribner, 2004.

Mr Fothergill's A–Z of Vegetables 2007. Kentford: Mr Fothergill's Seeds, 2007.

Pomodoroitaliano.it. Newsletter no. 3, September 2006. Available at http://www.pomodoroitaliano.it.

Snowden, Frank. *Violence and the Great Estates in Southern Italy: Apulia, 1900–1922.* Cambridge: Cambridge University Press, 1986.

"Speciale pomodoro." *Agrisole,* suppl. 3, January 20–26, 2006.

World Processing Tomato Council. "Tomato Processing in Italy." Available at www.wptc.to/pdf/italy.

Index

Numbers in italics refer to pages on which illustrations appear.

additives in, 110–12; agricultural subsidies for growing, 197–98, 200, 202–3, 214; ambivalence toward, 8–9; as aphrodisiac, 10–12; baked or grilled, 61, 153; bathtub cultivation of, 190–91; bottling of, 107–8, 167, 195; breeding of, 15, 86, 188–90, 201, 210; brought to Europe, ix, 2–3, 212, 212; canning of, 84–86, 89, 113, 115, 123, 125, 138, 140–42, 156, 157, 159, 163, 175, 186, 187, 188, 195, 201, 207, 213; classification of, 3–4, 12, 55; as "cold," 14, 15; color of, 2, 4, 15, 16, 26, 46, 47, 58, 62, 79, 91, 199, 211, 213; compared with eggplant, 1–2, 4, 5, 9, 11–12, 13, 14, 15, 26, 203; compared with tomatillo, 9–10, 21; as condiment, xi, 16, 49, 89, 99, 183, 211, 212; consumption of, x, xi, 55–56, 66, 71, 84, 95, 100, 115, 120, 143, 149, 177, 182–83, 187, 202, 212; cultivation of, 22, 51, 54, 65, 88, 105, 112–13, 140, 149, 183, 189, 190–91, 197–98, 201, 209; current production of (Italy), ix, 212; depictions of, 13, 13, 16–18, 17, 19, 22, 23, 24, 25, 45, 46, 46–47, 58, 87, 92, 199, 210; export of, 83, 84, 88, 109–13, 149, 164, 166, 212; fertilization of, 31, 65, 87, 88, 140, 188, 202, 207, 209; first reaction to, 1–2; flavor of, 15, 26, 49, 57, 58, 76, 79, 138, 199, 209, 210–11; as fruit, 5, 115, 210, 212; in Futurist cuisine, 152–54; genetic modification of, 201–2; greenhouse cultivation of, 51, 113, 209; as harmful, 11, 55, 183,

210; harvesting of, 65, 84, 115, 140, 186, 189, 197, 200, 201, 203, 205, 206; as healthful, ix, 55–56, 214; in kingdom of Naples, 65–66; as low-status plant, 26; as medicinal plant, 48; mythical associations of, 4; names for, 1–2, 3, 4, 9, 63; as ornamental plant, 16; in peasant economy, 59, 66; in pharmacopoeias, 12, 13, 48, 57; physicians' views of, 1–2, 5, 9, 10–11, 13, 14–16, 18, 48, 56, 67, 83, 84, 110, 190; as pill, 202; price of, 65, 140, 145, 200, 202, 210; processing of, 87–88, 113, 116–17, 136–40, 172, 173, 187, 201, 202, 204, 211, 213; proverbs about, 30, 80, 89–90, 155; pulp of, 142; raw, 67, 115, 199; removing skins and seeds of, 55, 61, 62; in salads, 190, 211, 213; sale of, 59, 66, 81; shape of, 3, 14, 16, 46, 49, 58, 62, 65–66, 91–62, 114, 190, 199; as "solar," 157, 195, 204; sun-dried, 58–59, 62, 79, 107, 179, 185, 195, 213; as "superfood," 214; sweetness of, 211, 213; throwing of, 64; toxicity of, 26, 214; use of, by Aztecs, 3, 4, 10, 12, 14, 16; use of, by elites, 56; use of, by Maya, 2–3; wild, 2–3, 24; yields of, 88, 147, 186, 187, 196–97, 200. See also *astrattu*; concentrate, tomato; *conserva*; *passata*; paste, tomato; preserves, tomato; purée, tomato; sauces; *individual dishes*

VARIETIES OF: acidity of, 56, 84, 89, 189, 190, 210, 212–13; beefsteak, 114; 'Cal-j', 189; 'Calroma', 201; cherry, 22, 24, 57; 'Chico', 189;